The Land of the Hunger Artists

From the 1880s to the 1920s, hunger artists – professional fasters – lived on the fringes of public spectacle and academic experiment. Agustí Nieto-Galan presents the history of this phenomenon as both popular urban spectacle and subject of scientific study, showing how hunger artists acted as mediators between the human and the social body. Doctors, journalists, impresarios, artists, and others used them to reinforce their different philosophical views, scientific schools, political ideologies, cultural values, and professional interests. The hunger artists generated heated debates on objectivity and medical pluralism, as well as fierce struggles over authority, recognition, and prestige. In the book, set on the fringes of the freak show culture of the nineteenth century and the scientific study of physiology laboratories, Nieto-Galan explores the story of the public exhibition of hunger, emaciated bodies, and their enormous impact on the public sphere of their time.

Agustí Nieto-Galan is Professor of History of Science at the Universitat Autònoma de Barcelona (UAB), ICREA Acadèmia Fellow, and director of the Institut d'Història de la Ciència (iHC) at the UAB.

The Land of the Hunger Artists

Science, Spectacle, and Authority, c. 1880–1922

Agustí Nieto-Galan

Universitat Autònoma de Barcelona

Shaftesbury Road, Cambridge CB2 8EA, United Kingdom

One Liberty Plaza, 20th Floor, New York, NY 10006, USA

477 Williamstown Road, Port Melbourne, VIC 3207, Australia

314–321, 3rd Floor, Plot 3, Splendor Forum, Jasola District Centre, New Delhi – 110025, India

103 Penang Road, #05–06/07, Visioncrest Commercial, Singapore 238467

Cambridge University Press is part of Cambridge University Press & Assessment, a department of the University of Cambridge.

We share the University's mission to contribute to society through the pursuit of education, learning and research at the highest international levels of excellence.

www.cambridge.org
Information on this title: www.cambridge.org/9781009379588

DOI: 10.1017/9781009379540

First published 2024

A catalogue record for this publication is available from the British Library

A Cataloging-in-Publication data record for this book is available from the Library of Congress

ISBN 978-1-009-37958-8 Hardback
ISBN 978-1-009-37956-4 Paperback

In memory of my beloved father, Quintí (1928–2020)

The symptoms of inanition consist of a painful pulling of the stomach, insomnia, delirium with stupor. Gradually, breathing slows down, the pulse becomes weak, the body temperature drops. The muscular system becomes weak, the blood becomes dry and the red blood cells disorganised; the quantity of urine decreases gradually, the stools become insignificant, greenish liquid. The stomach and the intestine undergo shrinkage, which is very dangerous when the subject starts to eat again … The fat, an organic reserve, a very oxidisable substance, disappears first. Then the spleen, the pancreas, the liver, the muscles, all the organs where important nutritious provisions exist, diminish gradually, as the secretions become concentrated and acidic. The brain is the organ that loses the least weight, which explains the persistence, sometimes very long, of the intellectual faculties, floating, so to speak, in the midst of the general debacle of nutrition.

Dr Ernest Monin, 'Succi', *La revue des journaux et des livres*, 16 October 1886

The propriety of yielding to a morbid popular taste in promoting entertainments which involve risk to life and limb is a question into which we cannot enter here, but as it has been asserted that important scientific conclusions might be drawn from observations made during the fast, we are glad to have the opportunity of examining the results of the medical men who watched Succi. The faster himself has apparently suffered no injury, and both he and the place of entertainment where he exhibited himself are, if rumour speaks correctly, considerably enriched by the feat of endurance; but the physiologist and the physician ask whether any more enduring good will result from the experiment, from the point of view of the increase of scientific knowledge and of relief to human suffering … we find that the scientific results of the experiment are as meagre as they well could be.

British Medical Journal, 21 June 1890

Contents

Figures

Tables

Preface and Acknowledgements

In 2010, while exploring some primary sources for the writing of a paper on the 1888 Barcelona International Exhibition, I came across an intriguing sentence, in a marginal footnote, which stirred my curiosity. It mentioned the case of an Italian faster, named Succi, who stayed in the Conference Hall of the Science Pavilion for thirty days without eating, only drinking liquids. I subsequently went to the local press and found some articles that dismissed this hype.[1] Some journalists even considered it as potential fraud, an insult to the prestige of a 'cathedral of rational, objective knowledge' such as the science pavilion of a world's fair. They denounced the incompatibility of such a peculiar performance with the scientific achievements exhibited in the venue.[2] At that time, I did not know who 'Succi' was, nor could I link such an odd event to other displays and performances that took place during the exhibition. Why did that mysterious 'charlatan' go thirty days without food in a science pavilion at an international exhibition? Was he part of the freak shows, panoramas, electrical spectacles, and animal parades that arrived in the city, as in many other urban contexts of the period, or was Succi an object of real interest for scientific study? These questions became the main driving force behind the writing of this book.

In 2012, after publishing my paper on the 1888 Barcelona International Exhibition,[3] I planned a research stay in Oxford, where my reading at the Bodleian Library provided me with further information about Succi and many of his colleagues, known as 'hunger artists'. I discovered that, from the 1880s to the 1920s – during the 'fasting for pay' period, as literature scholar Breon Mitchell described it – these intriguing characters performed an amazing number of public fasts throughout Europe and the Americas.[4] I also realised that, in 1922, Franz Kafka wrote the short story 'Hungerkünstler' ('A Hunger Artist') as a contemporary witness of this kind of public show.

Literature scholars have discussed at length Kafka's story, which epitomised a symbolic end to the profession of hunger artist, once entertainment tools for the urban masses substantially evolved when compared with the freak show tradition of the last decades of the nineteenth century.[5] In the text, Kafka crudely describes the loneliness and progressive marginalisation of an artist, a hunger artist, whose exhibition in a cage seemed to have lost the interest of the audience to the point that the circus impresario replaced him with a black panther in order to attract new visitors. Kafka's hunger artist was a loser, someone exhausted and depressed at the bottom of his cage, unable to keep the attention of the audience. With this tale, Kafka reflected on the fragility of the artist in general, and of hunger artists in particular, once the public interest in the performance declined. Some Kafka scholars even consider that Succi himself could have been the real character described in the story.[6]

Apart from Succi – an Italian of uncertain biography whose full name was Giovanni Succi (1851/3–1918)[7] – I soon came across the names of other hunger artists: Henry S. Tanner, Stefano Merlatti, Francesco Cetti, Breithaupt, Alexander Jacques, Agostino Levanzin, Victor Beauté, Riccardo Sacco, and many more, including numerous women such as Mollie Fancher, Claire de Serval, and Auguste Victoria Schenk, who appeared in the public sphere as 'fasting girls', but also as real professional fasters.[8] My task became even more complicated when trying to compare these hunger artists with the long tradition of religious fasts and the regular medical use of fasting for therapeutic reasons. Moreover, hunger strikes and force-feeding in prisons added more historical actors and political factors to consider, as well as cases of starvation due to accidents, natural catastrophes, and shipwrecks.[9] I soon realised that, despite being rooted in the old tradition of religious abstinence and mystical asceticism, public fasting progressively became a commodity, an amusement in the urban marketplace, but also a valuable subject for scientific study.[10]

Some years ago, focusing on Giovanni Succi, as a genuine representative of the hunger artist that Kafka described in his short story, I began to work on a particular episode, which later appeared in the journal *Social History of Medicine*. In 1888, in Florence, Succi fasted in public for thirty days, while the Accademia Medico Fisica of the city simultaneously subjected him to 'scientific' study under the supervision of the prestigious professor of physiology Luigi Luciani (1842–1919).[11] In addition, based on my work on nineteenth-century science popularisation and my later research on the urban history of science,[12] I then planned to extend the Florence case to other cities. In 2014, at the British Society for the History of Science (BSHS) Conference in St Andrews, I presented a preliminary version of a paper discussing a competition between Succi and another faster, Stefano Merlatti, which took place in Paris in 1886. Later, at the History of Science Society (HSS) Conference in Chicago, also in 2014,

I discussed the case of Succi in Barcelona, in 1888. In the first case, the Succi–Merlatti contest in Paris allowed me to join Irina Podgorny and Daniel Gethmann's project on 'scientific charlatans', which ended up in a special issue of *Science in Context*.[13] The two published articles and other presentations in seminars and conferences progressively led me to convince myself that it was worth revealing all the details of *the land of the hunger artists* that I present in this book.

Succi's life and impressive itinerary, from his early African trips to his numerous fasting tours, have given rise to an enormous amount of sound data. His close relationship with prestigious medical doctors who subjected him to experiments in the laboratory; his constant appearances in the press, in particular in the late 1880s and early 1890s; his agreements (and disagreements) with impresarios who managed his performances as a lucrative commodity; his controversial status as a potential charlatan – selling elixirs and never managing to be free of suspected fraud: all these made him a global celebrity. This book is therefore based mainly on the enormous primary sources that Succi generated in his time, but it is also an attempt to sketch the 'invisible college' that linked other hunger artists such as Tanner, Merlatti, and Cetti with Succi's 'heroic' accounts and adventures.

This is also the history of endless travellers. My historical account covers a very specific period, 1880–1922, from Dr Henry Samuel Tanner's fast in 1880 in New York to Succi's performance in 1890 in the same city, up to the symbolic year of 1922, when Franz Kafka published his short story. The history of hunger artists cannot be disentangled from its profound geographical nature, which implied moving from one city to another: performing fasts anywhere, but also walking, running, fencing, riding, climbing to the top of famous monuments and buildings, floating in real or tethered balloons, appearing on stages and in theatres, cinemas, and public lecture halls. Hunger artists' performances attracted considerable interest, and their shows were endlessly reported in generalist newspapers from all over the world, physiology textbooks, scientific and medical journals, popular science journals, photography and drawings, satirical publications, leaflets, reports of scientific societies, international exhibition guides, freak show archives, travellers' reports, and literary works. These vast primary sources have served as a powerful driving force behind the writing of this book, and have enabled me to build the historical account that follows. The way in which I have used and presented the secondary sources is obviously my own responsibility.

My work on the politics of science popularisation and its materialisation in urban contexts has been crucial for my study of hunger artists. I have benefited hugely from my ICREA Acadèmia research prize (Institució Catalana Recerca i Estudis Avançats) (2009 and 2018), awarded by the Catalan Government (Generalitat de Catalunya), and the support of the research group 2021 SGR 00015 'Science, Technology and Medicine in the Twentieth Century'. I also obtained funding from the Spanish Ministry of Science and Innovation through the project: 'Invisible Knowledge: The Politics of Censorship and Science Popularisation' PID2019-106743GB-C22.

In recent years, I have sketched *the land of the hunger artists* at numerous seminars and conferences: in Madrid, St Andrews, Chicago, Paris, Leicester, Barcelona, Milwaukee, and Madison.[14] Numerous friends and colleagues have discussed several aspects of the project, and their comments and criticism have been extremely helpful for the progress of my research. I am truly indebted to Oliver Hochadel, Daniele Cozzoli, Fernando Vidal, Richard Kremer, Elizabeth Neswald, Javier Moscoso, David Teira, Annette Mulberger, Mònica Balltondre, Alfons Zarzoso, Andrea Graus, Eduard Aibar, Jaume Sastre-Juan, Stefan Pohl, José Antonio Chamizo, Carl Nightingale, Melanie Beehler, Robert Fox, Anna Guagnini, Sybilla Nikolow, Charlotte Bigg, Andrée Bergeron, Eugenia Afinoguenova, and Silvia de Bianchi, among many others, for their suggestions and critical reading of my earlier papers on hunger artists, for their insightful comments, and for providing me with new primary and secondary sources.

Domenico Priori kindly sent me a very useful paper he published in 2017 on Succi and Luciani in Florence.[15] My conversations with Dr Robert Caner-Liese about Kafka's biography have been very helpful in linking hunger artists with contemporary literature and fiction. Although critical, the referees' comments on my European Research Council Advanced Grant (ERC-2014-ADG) application were very valuable in a long-term perspective. I am also grateful for the enthusiastic and constructive comments from the two anonymous referees. They generously read and wisely critiqued my manuscript, helping me to improve it substantially. My indebtedness also goes to Lucy Rhymer, Senior Commissioning Editor for History and Asian Studies at Cambridge University Press, for her meticulous and highly professional way of handling the whole process, since the submission of my first book proposal. Ailish Holly's stylistic corrections have turned my manuscript into a readable English text, and I am truly indebted to her for her patient work. Lisa Carter's revision of my images, tables, and maps, as Senior Content Manager at Cambridge University Press, was extremely useful. I am also very grateful to Karen Anderson, for her patient, meticulous copy-editing of the typescript.

I am also indebted to Pietro Mettelli at the Biblioteca Nazionale Centrale de Firenze (BNCF, Florence), who kindly provided me with the printed sources on Luigi Luciani and Giovanni Succi. The staff at the Biblioteca Marucelliana (Florence) were also very helpful in tracking down the remaining issues of Succi's periodical *Il Corriere Spiritico*. In Oxford, the staff of the Bodleian Library and the John Johnson Collection of Printed Ephemera were particularly useful. In Sheffield, the National Fairground and Circus Archive also provided useful data and images of hunger artists, living skeletons, and their shows. Access to digitised daily press has also been crucial for the gathering of data on hunger artists in Europe and the Americas. It is worth mentioning the Biblioteca Virtual de prensa histórica, the British Newspaper Archive, the Internet Archive, Newspapers.com, the Hemeroteca Arxiu Històric de la Ciutat (Barcelona), the Hemeroteca Nacional Digital de México (HNDM), the Österreichische Nationalbibliothek, Vienna, and Gallica (Bibliothèque Nationale de France) for allowing free access to very valuable files.

I wrote a large proportion of this book during the shocking and painful time of the COVID-19 pandemic, which brought distress and anxiety to family members, friends, and colleagues. In October 2020, COVID-19 caused my father's death, and I want to dedicate this work to his memory and to his unwavering support for my career.

Concepts such as *hunger*, *fasting*, *inanition*, and *starvation* are part of the present public sphere. Their historical dimension might interest a broad audience, from experts in nutrition and medical doctors to historians, humanists, social scientists, and journalists, keen to assess the cultural meanings of the apparently simple, routine mechanism of ingesting food (or refusing to do so). I hope the fasts that follow will stimulate the intellectual 'hunger' of a wide range of readers.

Introduction

In the autumn of 1886, a 'fasting mania' (*jeûnomanie*) invaded the city of Paris. Public lectures, newspaper articles, posters, photographs, caricatures, and engravings, but also scientific journals, academic meetings, and medical boards, all reflected the excitement about the presence in the capital of France of Giovanni Succi, the famous Italian hunger artist and professional faster. On this occasion, having already acquired very respectable credentials in previous fasts, Succi would compete with a young Italian painter named Stefano Merlatti on the capacity of both men to resist inanition for a long period of time. Merlatti arrived in Paris in early October 1886 to fast at the Salle du Zodiaque, on the first floor of the Grand Hôtel de Paris, next to Place de l'Opéra.[1] In early November, Succi was in Paris ready to fast at the Cercle de la Presse (rue Le Peletier).[2] As a standard procedure to avoid potential fraud, both fasts were supervised by medical boards, which included professional physicians, but also science writers, journalists, and medical students.

Succi and Merlatti soon became celebrities in the Parisian public sphere and regularly appeared in the press. The limits of human resistance to inanition, the multiple and often opposing scientific explanations of their resistance to hunger, the controversial biographies of both men, the composition of the liquids they ingested, and the suffering of their bodies in the absence of food all became trending topics that transcended academic boundaries. *Controversy, hype, fraud,* and *scientific authority* seemed inevitably cast together in a never-ending debate on the virtues and dangers of depriving humans of their regular ingestion of food. The impulsive young Merlatti took great risks in his long period of fasting, close to fifty days, at a time when contemporary medical treatises advised not going beyond forty days. His terrible pain shocked Parisian public opinion, while Succi, a more experienced faster, stopped on the thirtieth day, to preserve energy for future performances.

Caricaturist and science fiction writer Albert Robida (1848–1926) depicted this public excitement in his satirical journal *La Caricature*. In an imaginary banquet in protest against the 'fasting mania', he described the nature and dangers of Succi's and Merlatti's performances with the following words:

Oh, unhappy Italy! You gave us delicious mortadella and sweet macaroni, and now you send to us bitterness with Succi and Merlatti ... If these fasters' fatal doctrines spread, sirs, what about the culinary art, the principle of all things, the father of all arts. It will be the end, a man stifling all aspirations towards the good and the better will not be a man anymore. He will not be a vegetable either, since vegetables eat, but a simple mineral! Let us fight with all the strength of our stomachs, let us organise everywhere committees of resistance. Let us swear we will never quit each other before full indigestion.[3]

Likewise, popular excitement and uneasiness about the presence of the two hunger artists jeopardised the reputation of Parisian celebrities such as actors Benoît-Constant Coquelin – known as Coquelin ainé – and the famous Sarah Bernhardt. Both struggled to attract the attention of the urban audiences, who, in their view, were paying too much heed to the fasters. The press ironically reported that, due to Succi and Merlatti's success, Coquelin and Bernhardt also attempted to 'fast' for thirty days, avoiding any public appearance. The satire went even further when journalists figured out the 'medical' measurements of both artists, which were full of references to art and theatre, as an amusing imitation of Succi and Merlatti's physiological signs. As reported in *Le Figaro*:

In view of the success of Succi and Merlatti, Sarah Bernhardt and Coquelin, who do not shy away from any obstacle, have committed to fast for thirty days each. This is no ordinary fast. Instead of depriving themselves of food, these two great artists deprive themselves of advertising. They promise to remain for one month without being mentioned in any newspaper. Eight days have already passed since Coquelin remained immobile under the implacable surveillance of the severe Sarcey.[4] As for Sarah, she remains concentrated and silent in the hands of Mr Sardou and Mr Duquesnel.[5] On the ninth day, Sarah started to quiver ... An advert for tomorrow's *L'Evènement* is feared. Coquelin is holding out. It is true that he has been promised that he will be given an award on 14 July. Sarah Bernhardt too. *Bulletin on Coquelin:* Pulse: 114 pulsations; Dynamometer: resisted a Baour Lormian tragedy.[6] Breath: pleasant. Drank a glass of water which he immediately vomited. *Bulletin on Sarah Bernhardt:* Pulse: 3 pulsations; Dynamometer: Crushing of Ms Noirmont with one punch; Breath: perfumed; Weight: 0.5 centigrams. Drank a glass of Hunyadi Janos water,[7] which she involuntarily vomited.[8]

Regardless of who actually won the 1886 Succi–Merlatti contest in Paris, the enormous mobilisation that the apparently banal, charlatanesque show provoked among their contemporaries stirs the historian's curiosity. The two men's explorations of the limit of human resistance to inanition resulted in a cacophony of voices, which became a general pattern of hunger artists' fasts in different places in the last decades of the nineteenth century.

In 1888, just two years after the Succi–Merlatti contest, Dr Angiolo Filippi,[9] a distinguished member of the Florence medical community and editor-in-chief of the journal of the local Accademia Medico Fisica, *Lo sperimentale*, expressed his own doubts and worries about the presence of the famous Succi in the city and about the experimental study of his fast, which his colleague, Dr Luigi Luciani, was attempting to carry out in his laboratory. In fact, Filippi had doubts about considering the news about the visiting celebrity as a serious fact (*fatto serio*) or as worthless gossip (*sciocchezza curiosa*); he did not know whether to focus on Succi as a true hunger artist or as a professional faster (*digiunatore*), or even whether to dismiss him as a phoney (*mistificatore*). As a reflection of the blurred boundaries between health and insanity, between normal behaviour and ill manners, which tinged the credibility of the fasting performances, Filippi regarded Succi as someone in between a man of good sense (*uomo di senso*) and a man with an odd, queer brain (*cervello bolzano*). He wanted to know if all the effort invested by doctors at his academy would indeed have a useful application (*applicazione utile*) or be just a vain attempt. Filippi wondered to what extent the study of hunger artists actually contributed to actual improvements in medical knowledge – or perhaps these men were just useless charlatans, agents of fraud, of no scientific interest.[10] Filippi's reflections in Florence in 1888 are a source of inspiration for this book and for the pages that follow in this Introduction. His words brought to the fore crucial issues such as *trust, fraud, authority, spectacle*, and *heterodoxy*, which seem to be intrinsic parts of the slippery nature of the professional fasters. These late nineteenth-century concepts now re-emerge in our present as powerful analytical categories.

Hunger artists have always been on the fringes of historical research from different perspectives; they have never occupied the centre of gravity in the constellation of studies on food, hunger, and appetite. Although there is some historical research on professional fasters within general narratives of hunger and of self-starvation, we lack a full, comprehensive approach, which could provide new, previously unknown, reliable data, framed in an updated historiography.[11] Moreover, hunger artists appear as marginal actors in histories of nutrition and metabolism,[12] human experimentation,[13] and appetite and its appropriation by different medical schools,[14] as well as in cultural histories of food.[15] There are also occasional references to hunger artists in urban histories and in the cultural representation of cities (from freak shows to international exhibitions, museums, and public parks).[16] Fasting has also been linked to illnesses such as anorexia nervosa, as the pathological rejection of food for fear of gaining weight.[17] To the numerous ways hunger has already been addressed culturally,[18] in this book I add issues such as science (medicine), hype, spectacle and the marketplace, and their continuous mutual interactions.[19]

Food histories provide *longue durée* overviews, specific histories of foodstuffs or food events (sugar, cod, breakfast).[20] Also, they have addressed the therapeutic role of food (together with diet and fasting) in the domestic sphere throughout the nineteenth century.[21] Nevertheless, there is still a long way to go to integrate food history into a broader cultural history, in which hunger artists, in particular, can have their own place. In a recent volume of the journal *Osiris*, 'Food Matters', Emma Spary and Ayna Zilberstein stress the need to broaden our approach to food as a polyhedral object of historical research, which should include recent trends such as spatiality, political economy, globalisation, translation, and gender.[22] They suggest de-centring traditional cultural histories of food, beyond medical, scientific discourses. In a similar vein, Corinna Treitel, in her study of the nature of nutritional modernity in nineteenth-century Germany, emphasises that a historical approach to food needs to include a plurality of actors who played their role in the co-construction of habits and practices between the scientific, political elites and their civic, popular counterparts.[23] It is precisely under the banner of this broad, recent analytical framework that this book places hunger artists, also as multi-faceted objects of historical research, in the last decades of the nineteenth century.

Although some contemporary witnesses such as the prestigious French physiologist and Nobel laureate in Medicine or Physiology in 1913, Charles Richet, classified fasts into three separate categories: *jeûne experimental, jeûne charlatanesque, jeûne forcé* (experimental fasting, charlatan-like fasting, forced fasting),[24] the aim of this book is to explore the fluid boundaries between the experimental, the charlatanesque, and the forced nature of the events, and their place in late nineteenth-century culture. For this purpose, I shall describe the main features of *the land of the hunger artists*, a 'strange' territory, which probably does not fit into national containers, standard biographies, narratives of scientific progress, or the traditional approaches to performances and experiments. This is the history of a group of fasters and of the multiple ways in which their audiences appropriated them. Doctors, journalists, popularisers, politicians, impresarios, artists, priests, *gens du monde*, and more used hunger artists to reinforce their different – often opposing – philosophical views, scientific schools, political ideologies, cultural values, and professional interests. Fasters and their audiences, all of them worthy citizens of that land, contributed to the creation of an immense cacophony of public discourses, which poses serious obstacles for the historical reconstruction of the past but, at the same time, offers a rich cultural heritage to be preserved and studied.

From the seminal public fast of Henry S. Tanner in New York in 1880 to the publication of Kafka's short tale in 1922, hunger artists lived on the fringes of public spectacle and academic experiment, and acted as *mediators* between the human body and the social body.[25] Scientific controversies, opposing medical doctrines, and frequent suspicion of fraud apparently brought confusion and uneasiness, but at the same time they acted as epistemological challenges, as active driving forces for the emergence of new accounts of what actually happened inside and outside the fasters' cages.[26] Hunger artists therefore help us to assess how their expert audiences appropriated them, reported their own experiences, and contributed to the knowledge on hunger and inanition of their time.

At the intersection of the social, political, cultural, and scientific realms, this is therefore a history of an embodiment of connections, one that brings to the fore unexpected mechanisms of co-production of knowledge, endless debates on objectivity and medical pluralism, and fierce struggles for authority, recognition, and prestige on elusive subjects such as hunger, fasting, inanition, and starvation. Since scientific controversies about hunger and inanition were not limited to academic circles, they spread throughout an urban public arena that acted as a useful battlefield in that struggle for authority.

Co-production

In *the land of the hunger artists* some saw an excellent opportunity to gain public visibility, but others feared losing their professional reputation through risky alliances with questionable individuals. French populariser Wilfrid de Fonvielle (1824–1914) considered fasting performances, for example, to be a real threat,[27] an unacceptable pseudo-science, a moral calamity permitted by the indifference of the Parisian medical profession.[28] At the core of the Succi–Merlatti contest, Fonvielle warned of the risk of accepting hunger artists' spectacles as a 'scientific' study; he considered them as successors of Franz-Anton Mesmer – the Austrian doctor who 'mesmerised' Parisian audiences in the late eighteenth century with animal magnetism[29] – and of Allan Kardec – the founding father of spiritism.[30] Fonvielle bitterly questioned how prestigious scientific institutions made use of these controversial cases for their own professional interest.[31] In his view, it was not worth taking the risk of playing with potential frauds and with the limits of human life.[32] However, reluctance coexisted with hopes for scientific and professional improvement. The scientific culture of experimental physiology, which had flourished some decades before as a modern, ambitious field, acted as a supposedly reliable referee of the pretensions and exaggerations of those potential charlatans. Likewise, when compared with animals, the emaciated bodies could be valuable objects of experimentation on humans and could perhaps open up new research paths. Again, inspired by Filippi's scepticism and fascination, we can take hunger artists as a challenge to write a history that unveils a subtle co-production between science, hype, and fraud, between orthodox scientific authority and heterodox dissent, between academic expertise and popular commotion. This is a slippery zone, in which hunger artists and their publics, all the plethora of actors that surrounded these peculiar characters, actually co-produced knowledge on hunger and inanition. As American physiologist Sergius Morgulis discussed in 1923, just at the end of our period of study: 'The practical value of inanition will never be fully utilized until both laymen and the medical profession lose their instinctive fear of fasting. The experiences of recent years, which through the medium of the press have reached a large audience, will in the course of time alleviate the entirely unjustifiable fear of abstinence from food for longer periods.'[33]

From Richard Altick's *The Shows of London* to Iwan Rhys Morus's *Frankenstein's Children*, much has been written on the epistemological value of the spectacular in the nineteenth century, as part of the knowledge construction in public lectures, experiments, and performances.[34] Were those emaciated, hungry creatures comparable to exotic animals and plants exhibited in botanical and zoological gardens? Did they have something in common with the excitement provoked by machine displays in international exhibitions? Did hunger artists hold something similar to the fascination of dioramas and panoramas, magic lanterns, and stereoscopes? Or perhaps they fitted better in the context of freak shows? Can hunger artists be comparable to 'modern' spectacles such as photography, electric lighting, or the emergence of cinema at the end of the century, or were they perhaps just a relic of a past culture of spectacle, which progressively faded away around 1900? These are crucial questions to address the co-production between science – knowledge in a broad sense – spectacle, and hype in the chapters that follow.

In the crude exhibition of their emaciated bodies, hunger artists acted on the fringes of the freak show culture of the nineteenth century, often sharing the stage with impressive oddities, a performance apparently incompatible with their scientific study, in particular in physiology laboratories.[35] In the tension between science and hype, they had a lot in common with other freak creatures such as living skeletons and other popular performers of the social art of exhibiting monstrous bodies in itinerant fairs and dime museums.[36] Names such as Claude-Ambroise Seurat, an endless traveller and object of medical interest who inspired Francisco de Goya in his paintings;[37] Dominique Castagna, known as the 'mummy man'; George Prise, who acquired great popularity also in the 1880s;[38] John William Coffey, the 'Ohio skeleton' who was exhibited in American dime museums; Isaac Sprague, known as the 'Original Thin Man', among many others: all exhibited their bodies for the entertainment of the urban masses.[39] Similarly, the impressive number of visitors to the Paris morgue added more theatricality to the exhibition – in this case, corpses – for the curiosity and excitement of the public.[40] From its opening to the public in 1867 to its closure in 1907, the morgue became another public spectacle of exhibited corpses. It also vanished at the beginning of the new century, just at the end of the golden period of hunger artists and not long before the appearance of cinema, as a new entertainment for the masses, which progressively changed the rules of the game. Fasters, all sorts of freaks, living skeletons, corpses, and other oddities shared a common public culture, with no clear boundaries between science and hype.

Hunger artists can also be easily associated with the culture of charlatanry.[41] The nineteenth century and its global trends – in particular in the telegraphic era of the last decades of the century – witnessed a huge increase in the number of these controversial characters, who originated in early modern times, even the Middle Ages. Irina Podgorny and Daniel Gethmann have described in detail the typology of the charlatans that could easily include our hunger artists.[42] They all had an itinerant existence, dealing with exotic objects and remedies, organising shows and exhibitions, and performing miraculous healings. They usually crossed the boundaries between popular and learned culture, rich and poor audiences, women and men. They were equally celebrated and opposed by the learned elites of the time (natural philosophers, physicians, professional scientists), and adapted themselves to local contingencies. Charlatans used to experience exotic travels, a sort of grand tour that initiated them in the arts of their metier and contributed towards their public reputation. They became famous characters in the daily press and faced frequent accusations of fraud, often with unclear conclusions and never-ending controversies. Controversy, uncertainty, scandal, and even the risk of playing with the limits of resistance of the human body were appealing advertisements for potential customers who stood ready to buy the new commodity. Hunger artists must therefore be analysed in the charlatanesque culture of the time, but also in the context of entertainment and spectacle in late nineteenth-century cities, and within the medical challenges of contemporary physiology laboratories and scientific academies. It is only in that continuum of co-production that we can assess the epistemological value of their impact.

Physiologies

Although nineteenth-century physiology – the branch of medicine that studies the functioning of the body – was never unified and remained flexible, changing boundaries for decades, its historiography subtly carries a narrative of 'progress' in which laboratories, instruments, and quantification processes are probably over-represented.[43] In fact, intricate histories of the different schools of physiology, of plural views of the body, of the tensions between animal and human experiments, and of the epistemological power of places beyond the hospital and the laboratory can all contribute to a new history of physiology, in which the creation of new medical knowledge is embedded in a broader political, economic, cultural network.[44] Ciphers, graphs, and tables to study *fasting* (abstinence from all kinds of food), *inanition* (exhaustion – physical, mental, spiritual – caused by lack of nourishment), and *starvation* (suffering caused by lack of food, which leads to death)[45] coexisted with vitalist, holistic, spiritualist explanations in permanent tension with materialistic, mechanical models and sparked frequent controversy, speculation, and debate.[46]

As this book tries to exemplify, future research will find room between an over-optimistic history of modernisation, which presents an experimental physiology that can overcome all sorts of physical, moral, and social diseases and put 'unscientific charlatans' aside, and an over-pessimistic Foucauldian history of discipline, social control, and bio-power.[47] The study of hunger artists can therefore contribute to taming the polarisation between these two extremes. Professional fasters fit very well in the context of a historiography of the medical practice and their relationship with professional doctors for their mutual benefit. In fact, the fasters can be considered as a new kind of epistemologically relevant patient, adding reliable new data and cases to the practical turn in the history of medicine in recent decades.[48] They contribute to reinforcing the study of a nineteenth-century physiology that, as Richard Kremer pointed out, 'became not simply a special case of physics or chemistry but rather a science whose instrumental norms are inseparable from its public and private ethics'.[49] Also dealing with complexities and controversies, the study of local contingencies in medical societies and of local biographies of doctors, but also of new approaches to emerging specialities, has gained influence in the map of historical research. Beyond German and

French models, other countries' research schools and unknown figures will also have their place in future narratives.[50] As Natalie Zemon Davis pointed out, this is a history which is not simply told from the perspective of the elites, but rather aims to introduce plural voices.[51]

As David Livingstone discussed some years ago, particular places are active agents of knowledge co-production. Place matters as much as practitioners, since science exists in countless places beyond schools and laboratories.[52] As discussed in several chapters of this book, the spatial turn can also be applied to hunger artists and their relation with the physiology of their time. Geographically speaking, they embarked on itinerant journeys, while, spatially speaking, they performed in a great variety of places.[53] Many of them populated laboratories and scientific societies, but also sealed cages, theatres, museums, cultural institutions, international exhibitions, public parks, and urban streets. In addition, hunger artists' performances also fit very well within a new urban history of science,[54] which views the city as more than a simple container of scientific practices, but additionally as an active agent to shape expertise and the scientific culture of citizens' everyday life. Apparently, hunger artists' performances were carefully enclosed in cages and obsessively supervised by doctors and all sorts of overseers, even sealed in the sophisticated respiration calorimeters of prestigious physiology laboratories. However, at the same time, their social legitimation, their success as commodities in the marketplace, strongly depended on their social relations in different places outside the cage, and on their capacity to parade along streets, avenues, and bridges, and through all sorts of urban contexts.

Throughout the nineteenth century, hunger artists also became a source of inspiration for eclectic medicine. In the American context, for example, Henry Tanner's professional profile fitted very well in an eclectic school that opposed academic orthodoxy and chemical drugs and defended plant remedies, naturistic practices, the use of Native American medicine – as a sign of independence from the European schools – and a holistic understanding of the body. Eclectic doctors also opposed nineteenth-century mechanical, electric technologies, mass consumption, and industry.[55] In the same way that religion – in progressive America – reacted by reshaping spiritual experiences and rituals to offer tangible products to their followers, heterodox practices such as spiritism, mesmerism, homeopathy, hydropathy, and occultism also struggled to make their own space. In this context, individual, eclectic, personal credos – such as those of the hunger artists – gained prestige and influence and drew the attention of medical doctors and professional scientists, but also journalists, members of the clergy, philosophers, and literary writers.[56]

It was precisely in this atmosphere of medical heterodoxy that hunger artists could find their place in a crowded public sphere full of creators of different kinds of 'objectivity'.[57] Similar to how Kafka described the decadence of hunger artists in his 1922 short story, historian John S. Haller describes the decadence of the eclectics in the 1930s, since they could not free themselves from plant remedies or professionally face the new challenges of laboratory medicine and the germ theory.[58] In fact, the time of decadence of the eclectics seems to coincide with that of hunger artists and again can help us to revisit standard accounts of 'modernity'. The already mentioned optimistic narrative of a physiology based on laboratories, instruments, new specialities, and a successful medical professionalisation was offset by heterodox, eclectic practices, holistic resistance to reductionist, mechanical understandings of the human body, household health traditions, self-surveillance, self-discipline, and self-healing practices.[59]

In their time, hunger artists also contributed to revisiting the role of human beings as objects of scientific experimentation in relation to animals and challenged preconceived views of a 'standard', 'objective' experimental physiology.[60] Contemporary reluctance to engage in or allow animal torture and vivisection, and concerns about the ethical limitations of experimentation with prisoners, orphans, prostitutes, slaves, and the mentally ill, left room for hunger artists as a 'cleaner' option for the scientific study of the process of prolonged fasting. Experimentation with hunger artists also provided new psychological, non-strictly materialist data on the process of inanition which were hard to obtain with animals. In addition, fasting opened a door to self-experimentation,[61] in particular among medical students and doctors, such as Tanner's pioneering experience. In that context, it is obvious that hunger artists fit well in a history of nineteenth-century 'physiologies', in their many approaches.

Authority

We could approach hunger artists' stories as just another episode of freak shows and exhibition practices in the urban context of around 1900. We could also take them simply as cases of medical controversy and physiological debates, in a period of wide medical heterodoxy and the emergence of new medical specialities. Hunger artists could also be taken as an additional example of the political tensions around nutrition, minimum calorie standards, and the workforce as relevant indicators of nineteenth-century 'modernity'. Nevertheless, overlapping the different thematic chapters, this book aims to provide the reader with a unifying account, a continuum between the cultures of science and spectacle at the turn of the twentieth century.

The book provides an embodiment of connections (scientific, social, cultural, political), even if they often were not made explicit by contemporary actors themselves. As reflected in the chapters that follow, this is a history of the search for *authority, recognition*, and *prestige* through travels, performances, experiments, disciplinary boundaries, commodities, and exhibitionary complexes. They provide new and (I hope) convincing arguments to position hunger artists as relevant historical actors of late nineteenth-century culture and as crucial epistemological objects for the study of human fasting, inanition, and starvation in their time. Food and hunger, and hunger artists in our case, bring us to a terrain full of ambiguities, to a continuum of spaces, practices, and public discourses, which are hard to classify in separate spheres, and which I have tried to address in several layers of analysis as reflected in the chapters of this book.[62]

After this Introduction, Chapter 1, 'Geographies', describes how hunger artists' itinerancy became a key factor to explain their growing reputation and prestige in the period under study. It discusses how the geographical turn can be applied to the analysis of the tension between local and global events, to the spatial dimension of their practices, and to their professional status as itinerant travellers. The chapter describes Giovanni Succi's trips to Africa, Europe, and the Americas as a paradigm of the global dimension of a professional faster, and the way that colonial, commercial factors acted as preconditions for the later itinerant nature of the artist. The chapter also discusses the synchronic nature of the public fasts as described in the daily press on a global scale. Like natural catastrophes, wars, and accidents, the long fasts were immediately reported to urban readers worldwide and contributed to the emergence of new global publics. In addition, the geographical dimension of hunger artists also includes the dynamism of medical networks and research schools that shared results and experiments on fasting and added new nodes to the global network. Similarly, show business impresarios accompanied hunger artists and reinforced the itinerant nature of the metier.

Chapter 2, 'Performances', describes some of the most famous fasting practices, places, and institutions that welcomed hunger artists, from sealed cages to open-air shows, from Succi's impressive public fasting at the freak atmosphere of the Royal Aquarium London to his parade across Brooklyn Bridge in New York in 1890. Guided by Kafka's 'Hungerkünstler', which reflected on the influence of the audience in the success or failure of fasting performances, this chapter describes a general typology of performances of public fasting and their implications in the epistemology of the art. It depicts hunger artists' fencing, riding, climbing, and ballooning in specific urban contexts and their influence on contemporary knowledge on fasting, inanition, and starvation. The chapter also analyses how the varied audiences appropriated hunger artists and created a commotion in which different and opposing interpretations of the facts coexisted in the public sphere, for example, concerning the risk of exceeding thirty days of fasting, and the ideal method to tackle fraud and provide 'objectivity'. It also points out the controversial views on the scientific value of public fasting performances in medical journals and in the daily press.

Chapter 3, 'Experiments', analyses how hunger artists reinforced, or sometimes questioned, the authority of a scientific experiment, in particular the value of three key spaces: the laboratory, the bed, and the bench in experimental physiology. Through several cases of public fasting, it revisits traditional ways of assessing the epistemology of the experiment and the reliability of the new instruments of physiology that spread through the medical geography of the second half of the nineteenth century. The chapter also discusses the tension between animal experimentation and human tests in processes of inanition, and the complex relationship between doctors and hunger artists as objects of scientific research through the history of partners such as Drs Ernest Monin and Philippe Maréchal and hunger artist Stefano Merlatti; Dr Luigi Luciani and hunger artist Giovanni Succi; Drs Curt Lehmann and Nathan Zuntz and hunger artists Francesco Cetti and Breithaupt; and Dr Francis Gano Benedict and faster Agostino Levanzin. It reflects how doctors' scientific interest in the study of the human body in a process of prolonged fasting also served hunger artists' interests in public legitimation and social prestige, but often created tensions between the experimenters and the strict conditions endured by the objects of that scientific experimentation. The chapter also reflects hunger artists' contribution to research on human metabolism, and the value of the study of the process of human inanition to quantify food intakes and outputs and of the internal consumption of the different tissues of the body. It also points out the doctors' limitations in their numerous attempts to draw a clear line between the 'inner', 'scientific' space of the laboratory and the respiration calorimeters and the 'outer' commotion of the public performances, with the 'sealed cage' being a sort of common place to blur that dichotomy.

Chapter 4, 'Spirits', discusses the limitation of strict materialist explanations of the processes of inanition, and how hunger artists fitted into alternative vitalist, naturist, eclectic, spiritualist approaches, which at the time had numerous defenders. Succi, being the editor of the spiritist journal *Il Corriere Spiritico*, is a paradigmatic example of this trend. In that context of heterodoxy, hunger artists contributed towards opening the door to a psychological turn, to the progressive emergence of psychological explanations of voluntary hunger and resistance to inanition as a new field of scientific inquiry, often with a gender bias. The psychological turn gave new prominence to names such as Charles Richet and Hippolyte Bernheim, but it also had its roots in Luigi Luciani's non-strictly materialist explanation of Succi's resistance to inanition in Florence in 1888. The chapter brings to the fore the names of several women fasters, often treated as patients and pathologised as 'fasting girls', but in other cases appearing in the public sphere like other male professional fasters and following analogous performances. In that gendered psychological turn, terms such as *willpower, inner force, hypnosis,* and *insanity* progressively gained influence in explanations on the causes of resistance to prolonged fasting.

Chapter 5, 'Elixirs', discusses how hunger artists' fasts were associated with a huge range of drugs, liquors, and mineral waters, which provoked scientific controversies and public disagreements, but at the same time strengthened advertising campaigns in the medical market. The chapter also discusses the close link between hunger artists and homeopathic doctors in specific local contexts, in particular in the case of Succi's fast in Barcelona in 1888, and the analogies between fasting practices and homeopathic regimes. The ingestion of specific liquors – such as Succi's famous liquor – which supposedly helped the fasters to withstand the pain of hunger in the first days, never achieved consensus among analytical chemists and doctors, nor was there any agreement on their narcotic or nutritive nature. Equally, in the battle to draw boundaries between orthodox and heterodox science, the composition of different mineral waters was an extra tool for advertisements in which doctors and hunger artists became active, complementary agents of credibility. Again, issues of trade, fraud, and scientific objectivity resulted in controversy and frequent disagreements, but strengthened the promotion of hunger artists and their performances.

Chapter 6, 'Politics', describes how hunger artists became a trending issue for the morality and politics of industrial societies of the late nineteenth century. Their controversial performances challenged hygiene policies and issues of individual and social discipline as well as adding new factors to be considered by social reformers. Since hunger artists entered the popular culture of their time, their 'heroic' stories contributed to the debate on the possibility of human beings living with less food, so the social conditions of wellbeing and health, especially for the working class, could be revisited. Public fasting became a sort of physical prowess, a metaphor of self-discipline, a commodity to be bought and sold in the logic of the industrial capitalist society, at the same time as opening the door to a more popular, eclectic medicine that challenged academic authority and established power. Equally, despite their global nature, these popular, controversial performances became a tool of national pride or national humiliation. Hunger artists played a role in standardisation processes of the calories required to properly feed the citizens of the nation. They evoked the health, resilience, and discipline of the average citizen as key agents in the making of the modern nation, as a future, collective project.

Finally, the Conclusion presents the potential benefits of human fasting experiments and performances to 'advance' medical knowledge, as described by Dr Robert Alexander Gunn, a member of the supervisory board of Dr Henry Tanner's fast in New York in 1880, and one of his biographers. It is through the historical analysis of these 'benefits' that this final chapter discusses the actual epistemological value and historical agency of hunger artists, and summarises the main analytical frameworks of the book. This final chapter also recapitulates the main landmarks of the rich 'topography' of the land of the hunger artists from the long list of cities in which public fasts took place, to the variety of 'citizens' of the land who circulated across its vast territory and the places and material objects that brought doctors, fasters, impresarios, and global publics together. At the end of the Conclusion, I have added a short discussion about the reasons for the hunger artists' decline in the 1920s, in an endeavour to place Kafka's pessimistic view of the metier in a proper context.

The land of the hunger artists apparently replaced the old mystic, religious, ascetic forms of fasting with secular spectacles and urban commodities. However, as Kafka suggested in his 'Hungerkünstler', the 'artistic' nature of the faster was part of the artistic resistance against modern bureaucracy, division of labour, managerial strategies, urban transformations, and the commodification of social relations. Hunger artists were a good representation of some of the contradictions of the modernity of their time. On the one hand, they became another standardised product in the urban marketplace. But they occupied a blurred, uncharted territory in which concepts such as 'rationality', 'objectivity', and 'progress' were subjected to severe scrutiny. The blurred boundaries of their metier between science and spectacle, the permanent suspicion of fraud regarding their supposed charlatan-like nature, and the difficulties that academic science had to face to explain their peculiar behaviour placed them between order and chaos, localism and globality, authority and resistance.[63] This is the fascinating nature and the tremendous historiographical challenge that hunger artists represent.

1 Geographies

In 1882, the Italian geographer Federico Minutilli (1846–1906) lectured on the role of Africa in science, civilisation, and trade at the Geographic Society of Rome. Minutilli, a professor at the University of Pavia, had worked on cartography and demography, and in 1902 he published a widely read geography book, *La Tripolitania*, to popularise the name of Libya, one of the territories of the former Ottoman Tripolitania, as a symbolic precedent of the colonial domination in the Italian Libya.[1] In that context, Minutilli used the opportunity to present to the audience Giovanni Succi's trips and commercial endeavours. At the end of his talk, Minutilli invited Succi – who introduced himself as a 'respected explorer' – to say a few words.[2] Succi mentioned his acquaintance with Pierre Savorgnan de Brazza (1852–1905), the founder of Brazzaville, in the new Republic of Congo, who was also in contact with Henry Stanley; Succi seemed to have met Stanley during his African trips (Figure 1.1).[3] Both explorers, Brazza and Stanley, played a role in the Berlin Conference of 1885 for the colonial control of Congo. Brazza became governor of the French Congo in 1886, one of Succi's most successful years as a hunger artist.

Succi built up his public image as a traveller and adventurer, which accompanied him in public lectures, meetings in scientific societies, and advertisements that preceded his fasts. He placed himself in the Olympus of the great names of Italian geography and exploration, in the context of the colonial enterprise of a nation proud of its recent unification.[4] Succi's trips became powerful tools to gain reputation, social trust, and scientific authority. They acted as a sort of symbolic account via which he introduced himself in public and drew the interest of broad audiences. The nature of hunger artists seems therefore indissociable from geography. Their itinerant practices, their continuous travels, and their promises of miraculous cures and elixirs became an intrinsic aspect of the metier. In Succi's case, starting with an initial exotic experience in Africa, his journeys extended to Europe and the Americas. Obviously,

Figure 1.1. Giovanni Succi in Africa in the company of the English explorer Henry Stanley, c. 1880.

performances in cities such as Paris, London, and New York had more repercussions in the public sphere, but the accumulation of multiple micro-stories resulting from numerous trips became key milestones in the making of the uncharted territory and the design of the land of the hunger artists.

Public fasting became the target of newspapers and magazines seeking new global publics.[5] Benefiting from telegraphy and news agencies, the press synchronously reported hunger artists' fasts in similar terms to natural catastrophes such as earthquakes, volcanic eruptions, flooding, and technological accidents such as industrial explosions and railway crashes.[6] Even fraudulent performances and news about impostors – sometimes stealing the names of the great luminaries of the art of fasting – appeared in printed media across continents. The press reported their arrivals and departures into and out of every city, details of medical boards, local reactions, and displays. Anecdotes, caricatures, and medical concerns about these painful, disgusting experiences spread across the planet and contributed to the making of the land of hunger artists, which can be viewed as another example of the globalisation trends of time and space in the nineteenth century.[7]

The chapter therefore follows hunger artists in motion, through travels and networks, in local sites, cities, countries, and even continents, and through the news of their performances, which contributed – albeit with some controversy – to their professional reputation.[8] To this map of hunger artists' travels, and the synchronicity of the global news, one should add the circulation of doctors and impresarios, with their instruments, experiments, advertisements, and business plans, who accompanied the professional fasters across that land. In addition, the circulation of people, objects, and information through an international network of schools of physiology and nutrition added more density to the web. Francis Gano Benedict became, for example, a key player in the construction of a network of schools of physiology (and nutrition, in his case) that utilised hunger artists as objects of scientific inquiry and shared experimental results (see Chapter 3).[9] His case can be extended to other laboratories that constructed a geography of physiology, which, in our case, took hunger artists as relevant objects of study. It is therefore the dynamism of the art of fasting circulating through these networks that conferred a reputation, prestige, and authority to the metier, which deserves further analysis in the following sections.

Travels

In Rome in 1876, Succi met Amid-Mohamed-Abdallah, prince of Johanna (today known as Anjouan, an island in the Comoran archipelago), who became his main commercial contact in the years to follow.[10] Succi's trips to Africa started as a strictly commercial mission, but ended up being a more personal and emotional experience. His book *Commercio in Africa*, which appeared in Milan in 1881, was a summary of his early commercial missions and provided a detailed description of the African land he had explored in person.[11] In 1877, Succi began his travels from the Italian Adriatic coast to Constantinople, and then to East Africa, where he planned to spend three years around the Comoran archipelago, Madagascar, the sultanate of Zanzibar, and Mozambique. A fever in Dar es Salaam, supposedly caused by some plant extracts, apparently helped him to discover his innate resistance to inanition.[12] After days and days under the tropical sunlight with long daily journeys of almost 30 km on foot, he seemed to have spent eleven days without food, a seminal experience that would have a great impact on Succi's life some years later, once he was back in Europe (see Figure 1.1).

From the Cape of Good Hope, in a very weak state, Succi embarked on a trip to Buenos Aires and Montevideo to return from there to Italy in 1879. He then tried to fund his own commercial society for trade between Italy and the African regions he had contacted, but in the end he joined the Società Italiana di Commercio to open a commercial base in Zanzibar. Succi went back to Africa in 1880, to Zanzibar, the Comoran islands, and Madagascar, to develop his own trade project. He was particularly interested in the island of Johanna for its natural resources and its central position in the Comoran archipelago. In 1881, he reached a commercial agreement with the local authorities for imports and exports to Italy.[13] Succi stayed for a long time on the island of Johanna, which he described as: 'The island that more than any other, right from the first trip, attracted my attention, for the richness of its products and its central position in the Archipelago, was the island of Johanna, which today would acquire a vital importance for trade, having been able to obtain from that King full and perpetual exemption for all items of import and export'[14] (Figure 1.2).

Succi's reputation as a traveller, trader, and explorer rose even higher after some evidence of a meeting with Stanley. In an interview in the *Pall Mall Gazette*, some years later, Succi stated that: 'It was there that I fell in with Stanley, near the Voami River, where I was hippopotamus-shooting.'[15] Fiction apart, Succi's profile as a traveller and explorer accompanied him many years later, during his fasts. As in the case of Minutilli in Rome, as well as in other geographical societies, in 1886 in Paris Italian doctor Enrico Barberi-Borghini – a member of the medical board of Succi's previous fast in Milan the same year – introduced him to the Société de géographie.[16] In front of an audience of approximately 800 people, including a considerable number of medical doctors and journalists, Barberi-Borghini lectured on the most relevant cases of fasting in history, from ancient times to the present day.[17] Dr Prosper de Pietra Santa (1820–1898), a prominent figure in public health in France and the editor-in-chief of *Journal d'hygiène*,[18] chaired the session. Later, Barberi-Borghini introduced Succi as a successor to the 'famous' American doctor Henry Tanner. In 1876, Barberi-Borghini had already met Succi and noted his enthusiasm for travel and exploration. After a detailed account of Succi's African adventures and the discovery of his capacity to fast – a standard narrative that later appeared in the press on a global scale – Barberi-Borghini invited the French doctors in the audience to assess the fast. Succi used his epic trips to legitimise his extraordinary experiences. In fact, he later talked about how his African experience had opened the door to his heterodox practices

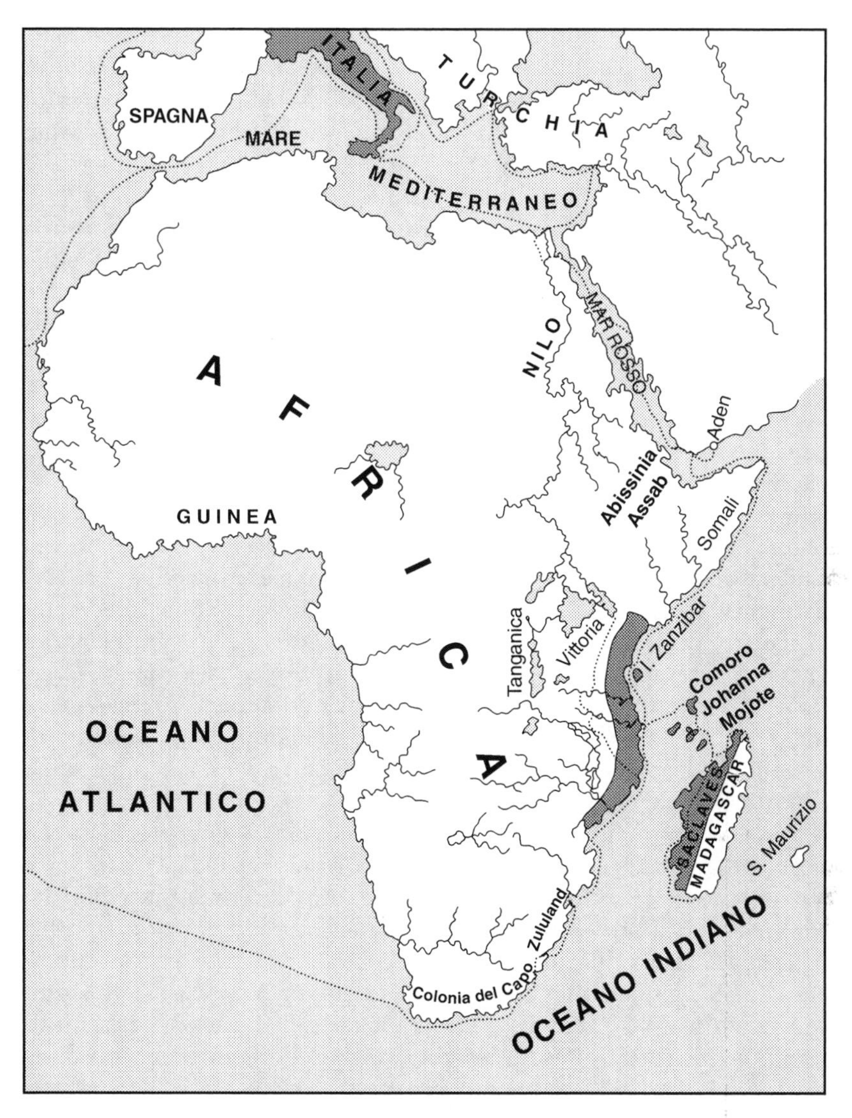

Figure 1.2. Giovanni Succi's commercial trips in Africa 1877–1881. Places where he aimed to open new trade appear in grey. He established a trade company and a general goods store on Johanna island. Discontinuous lines show Succi's trips in Africa in the period 1877–1881. In 1881, Succi established the central premises of the Agenzia della Società Italiana di Commercio coll'Africa in Zanzibar.

and perceptions of the body: 'It was in 1879, when ill with fever at Dar-esalaam [*sic*], on the African coast, that *I first discovered my capacity for abstinence.* I had fasted for several days, and instead of losing strength, I suddenly felt a new force awake within me, which the sustained exertion of my will seemed to increase.'[19]

Succi encountered some African plants with rare therapeutic properties and reported that he had met some 'spirits' who supposedly inspired his later crusade as a hunger artist (see Chapter 4). The myth of Succi's African voyages and their role in his transformation into a professional faster reached the European public. Doctors who studied Succi in his subsequent fasts stressed the importance of this exotic African experience. Dr Giovanni Chiverny, for instance, president of the medical board of Milan, pointed out that it was during Succi's travels through Africa that he discovered his exceptional skills when a fever weakened his liver and spleen, but he overcame it through days in inanition.[20] Exotic travels, a fictional narrative, rare therapeutic plants, mineral waters (see Chapter 5), spiritual experiences, and commercial, colonial endeavours, all combined in interviews, newspaper articles, and advertisements, constructed the public image of the exceptional character of the Italian faster Giovanni Succi, who quickly went from being an unknown character to a global celebrity. In Paris, caricaturist Albert Robida satirised Succi's epic accounts through the story of Luigi Macaroni,[21] a fictional character representing Succi and speaking French with a comical Italian accent, who had embarked on an exotic, disappointing adventure (Figure 1.3).

Succi was born in Cessenatico, near Forlì, in Italy, sometime between 1851 and 1853 (Figure 1.4). His father was a sailor who died when Succi was only fourteen. After some education, the young man worked as a bank clerk in Rome, but soon travelled abroad and embarked on his African adventure. Back in Rome, Succi performed an impressive number of public fasts, first in Europe and later in the Americas. With a firm conviction that his African travels had made him extremely resistant to inanition, he began to fast in his hometown of Forlì and later in Rome. It was in 1886 in Milan, mainly through the mediation of Dr Barberi-Borghini, that his performances began to acquire a public reputation in the national and international press. Also, that same year, in Paris, Succi faced the famous contest with faster Stefano Merlatti. Two years later, in 1888, in Florence, he entered Luigi Luciani's laboratory of physiology – one of the key episodes of the use of hunger artists' performance for medical purposes (see Chapter 3).

Succi later appeared at the Science Pavilion of the 1888 International Exhibition in Barcelona and then travelled to Madrid, Lisbon, Cádiz, and Seville. In 1889, he fasted again in Paris, at the 1889 International

AVENTURES DU CÉLÈBRE JEUNEUR LUIGI MACARONI, — par GEORGINA

ADVENTURES OF THE FAMOUS FASTER LUIGI MACARONI, by GEORGINA

Transported by the steamboat the *Méduse*, in one of the most remote parts of the ocean, Luigi Macaroni lets himself be carried away by the wind.

60th day. *Per Bacco*! I will prove that fasting and Hunyadi Janos water are the greatest conquests of science!

180th day. *Par la Madone*! I should have worn an overcoat!

275th day. The zephyrs have taken Mr Macaroni to the island of Thibia-grati-nophom inhabited by the tribe of the Bouff-lé-néh-soss-tomath, professional man-eaters.

The Bouff-lé-néh-soss-tomath are gourmands who look down on such a thin delight.

Macaroni is deeply humiliated by this.

Even the king of animals is disdainful. Macaroni is increasingly humiliated.

You don't want to eat? OK, then! I'll eat you!

365th day. Return to Paris. A lion apple purée! There you go!

Figure 1.3. 'Aventures du célèbre jeûneur Luigi Macaroni, par Georgina', *La Caricature*, 25 Dec. 1886, 421.

Figure 1.4. Giovanni Succi (photograph c. 1900).

Exhibition, and later in Rouen and Brussels. In 1890, he fasted at the Royal Aquarium of London (see Chapter 2).[22] That same year, Succi travelled to the United States and fasted in New York. In 1891, he visited the Chicago World's Fair, whereas in Boston he approached the freak culture of a dime museum. After returning to London in 1891, he continued his European tour: Naples (1892), Rome (1893), Berlin, Geneva, Budapest, and Marseille (all 1895), Vienna[23] and Zurich (both 1896), Verona, Florence, and Rome (all 1897), Genoa (1901), and Munich and Hamburg (both 1904). Finally, he embarked on a trip to Latin America to fast in Havana, Mexico City,[24] and Buenos Aires (all 1905).[25] Back in Europe, Succi fasted in Bologna in 1907 at the entrance of Cinematographo della Borsa, as a sign of the new urban entertainments that were replacing the old freak show context, probably anticipating its decline. All these fasts, and probably many more that have been poorly documented, turned out to be a lucrative business until Succi's popularity decreased in the early years of the twentieth century. There is evidence of some late fasts in this period in the company of a Madame Succi[26] but, like Kafka's hunger artists, his job seemed to have reached its end. Having disappeared without a trace in the ranks of vagrancy, Succi died in Rome in 1918.

Succi's intrinsic itinerant nature can also be observed in other hunger artists. Born in the early 1860s – probably in 1864 – in Mondovì (Piedmont), near Turin, Italy, faster Stefano Merlatti soon combined his calling as a painter with early fasting experiences at the Turin Academy of Fine Arts, when he was just twelve years old. In 1885, Merlatti had fasted for thirty-six days in London and had continued to work as a painter,[27] but he acquired a public reputation in Paris in 1886 under the auspices of Drs Philippe Maréchal and Ernest Monin.[28] In 1887, a Norwegian of Italian origin Francesco Cetti (1860–1925) fasted in Berlin for thirty days, surrounded by journalists, physicians (including Rudolf Virchow), and other watchers and onlookers (Figure 1.5).[29] Like Merlatti, Cetti was an artist, a musician in his case, and a famous aeronaut with very popular balloon ascensions, a fashionable practice at the time, and another way of exploring the human capacity to resist hostile atmospheric conditions.[30] Cetti was of Italian descent, but he grew up in Bergen where his father worked as an instrument maker. After studying theatre, he became a travelling musician and later a juggler. Cetti travelled by balloon through Norway and Denmark and later acquired his own hot-air balloon, *Christiania*. In 1892 he moved to Sweden, and after some journeys in Scandinavia, in 1897, Cetti went to Argentina to organise a balloon corps.[31] He gained visibility among the medical profession by being studied in 1888 by prestigious physiologists in Berlin,[32] alongside a man named Breithaupt, who was a journeyman shoemaker and professional faster (see Chapter 3).

Employed in silk printing in Dartford, England, and later in Épitaux, France, Frenchman Alexander Jacques miraculously survived the Franco-Prussian War, and 'fasted twenty-eight days, his only sustenance being his elixir, prepared from herbs';[33] he had supposedly inherited the recipe, designed to prevent starvation, from his grandmother.[34] After earlier fasts in Europe, in 1892, he crossed the Atlantic. Under the controversial title of 'Emigrants of Doubtful Utility', the press reported Alexander Jacques's arrival in the following terms: 'Another champion hunger artist arrived in New York on the steamer Gallia ... His name is Alexander Jacques and he speaks of Succi with fine gallic scorn. Medals on his breast testifying that he fasted fifty days at the Royal Aquarium London during the autumn of 1891, 42 days another time, 47 days in Christiania (now Oslo) and also in many more places.'[35] In fact, Jacques visited some of the key places in which hunger artists usually performed. He emulated Succi in the freak show

Figure 1.5. Francesco Cetti in a balloon ascension (n.d.).

atmosphere of the Royal Aquarium London (see Chapter 2) and also had shared experience of Christiania with Cetti.[36]

Other hunger artists even reached Australia. This was, for example, the case of Giuseppe Sacco-Homann, who, like other hunger artists of a similar name, adopted 'Sacco' as a phonetic alternative to 'Succi' (Figure 1.6). Of Hungarian origin, he fasted in Germany in the early 1890s and some years later in London, at the Royal Italian Circus (1906) and the Olympia (1907). In 1908, Sacco-Homann fasted at the Operetta House in Edinburgh and later at the King's Hall in Birmingham. He later travelled to Australia, to fast at the Melbourne Waxworks, and there is evidence of other fasts back in England in 1912.[37] In fact, that year, the press reported Sacco-Homann's fast in Ipswich in the following terms: 'Giuseppe Sacco-Homann, the fasting man, on Friday afternoon completed eleven days of his self-imposed fast in his quarters at Hyde Park Corner, Ipswich … The chief trouble has been the cold … Sacco-Homann explains that he has just come from Australia where it is much warmer than in this country. Besides, one feels the cold much more when

Figure 1.6. Giuseppe Sacco-Homann, 'The original world champion fasting man', 1900.

the food supplies are cut off. This Dr Tanner of the Twentieth Century has beaten all modern fasting records, having on three occasions accomplished the wonderful feat of fasting for fifty days and over.'[38]

Travelling became an intrinsic part of the hunger artists' metier, in tune with the charlatan-like profile of their professional strategy. Their inter-urban journeys acted as links between the local and the global, in a context in which news about the processes of their resistance to inanition, but also suspicion of fraud, quickly crossed continents and became an appealing topic for the global press. Hunger artists' travels also served as mythical heroic narratives, especially the initial ones – such as Succi's trips to Africa – which were often combined with entertaining accounts of their local problems and the contingencies they encountered in different cities and places. Travels became a key pillar for the construction of the interwoven net of the land of the hunger artists. With blurred boundaries between centres and peripheries, their journeys brought together different historical players who accompanied the artists. The moving, dynamic nature of the hunger artists' metier was closely linked to their need to build a blurred, but useful, public sphere to legitimise their profession. In that sense, the geographical dimension of hunger artists acquired an epistemological status. Contingencies and anecdotes in local contexts were quickly shared through the press at a global level, so they constructed a general pattern of the metier. They constituted a key pillar of a geography of knowledge which was very much in tune with contemporary global trends.

Synchronicity

Hunger artists were part of the global public of the late nineteenth century.[39] At a time when markets, commodities, and communications networks were considerably globalised, hunger artists became relevant actors in the micro-histories of the global.[40] From the early 1870s, information travelled at high speed, and apparently humble local events soon acquired international importance. In 1883, for example, news about the eruption of the Krakatoa volcano very quickly spread across the globe.[41] In fact, historians have described in detail how multiple exchanges of people, commodities, and information developed throughout the nineteenth century.[42] The formation of modern nation-states was counterbalanced by cosmopolitan, internationalist projects, world's fairs, and artificial languages, but also by the extensive use of telegraphy and transatlantic cables and, particularly in the second half of the century, by the emergence of a modern press that promoted debates among urban readers from all over the world, in particular in Europe (and its colonies) and the Americas.[43] The period of study of hunger artists roughly coincided with a period of impressive technological innovations in communications and trade.[44] These were the times of electric lighting and motors, internal combustion engines, typesetting and typewriting, cheap paper, parcel post services, global insurance movements, telegraphy, the telephone, and, finally, wireless communications. These were also the times of the emergence of global publics, of courts of global opinion which provided huge marketplaces and arenas in which hunger artists found their place.[45] In many cases, local events, details of a single fast in a specific local context, immediately occupied pages on international news in numerous newspapers. In other cases, simultaneous public fasts occurred at the same time in different cities and allowed witnesses, from doctors to journalists, to report details of the events synchronically.[46]

Tanner considered that his fast in New York became a 'Great American sensation, and his novel incidents were wired to the ends of the telegraphic world'.[47] Since 1880, news of Henry S. Tanner's fast had quickly

Figure 1.7. Henry S. Tanner after fasting for fourteen days at Clarendon Hall, New York City.

spread to the whole world and had become a topic to discuss, imitate, or reject, but never ignore (Figure 1.7). Curiously, Tanner became an icon for European hunger artists at a time in which the United States played a rather marginal role in relation to European science. His heterodox medicine and the heterodoxy of hunger artists themselves created a network which did not necessarily follow a standard pattern in terms of the centre/periphery of the scientific endeavour. In Paris, Robida's journal perceived the impact of Tanner in France as a sort of invasion of '*docteurs excentriques*', whom the medical establishment considered as suspicious but who in practice had a great effect on the popular medical culture.[48] In fact, suspicion of fraud during Tanner's 1880 fast quickly crossed the Atlantic. In September that year, the Spanish satirical journal *El Loro* published the rumour that Tanner had probably taken a Liebig meat extract during his fast; this very popular food complement had already become a commodity since it had been introduced to the marketplace after its industrial development by the Justus von Liebig vegetable and animal chemistry laboratory.[49] In the mid-1880s, the daily press regularly commented on the international impact (in Europe and throughout the world) of Tanner's fast and of hunger artists' performances as a whole.[50]

In 1886, just three years after the eruption of Krakatoa, three simultaneous fasts took place in different cities. That year can be considered as an *annus mirabilis* of the profession with a high density of events in the map of the land of the hunger artists. In Paris, Succi and Merlatti met to compete in the human resistance to inanition, while their health condition was reported daily in short notes in newspapers. At the same time, the *Argus*, a newspaper from Melbourne, provided details on Merlatti's biography, which were hard to find in the European press.[51] Also, in 1886, a hunger artist named Mr Simon fasted in Brussels. Under the supervision of a medical board, Simon wanted to attempt a fasting world record, but stopped after twelve days. He claimed he had humanitarian motivations for his fast, explaining that it was to serve as an example for miners who might be trapped in a collapsed mine, but could stay alive for a long time without food (see Chapter 6).[52] Simon's fast also resounded on the other side of the Atlantic; it was reported in the *Toronto Daily Mail* in the following terms: 'Simon, the man who is fasting in Brussels, has reached his seventh day in good spirits. He walks, plays the piano, and even dances. He feels well, although he has lost ten pounds in weight, and says that after the thirtieth day he will go to England to show by a sea trip that he is none the worse for fasting.'[53]

Also in 1886, Alexander Jacques wanted to compete with Merlatti and Succi and fasted for twenty-eight days in London.[54] In fact, the press also reported the synchronicity in the following terms: 'A Swiss named Simon, emulating Succi and Merlatti, began a fast [in] Brussels which is to last for 31 days. Alexander Jacques, a printer of stuffs, has undertaken a fast longer than Succi, who accepts his challenge. Dr Gubb and two medical students have agreed to watch night and day over the interesting experiment which commenced at Dieudorne's hotel, Ryder-street, St. James.'[55] A French dictionary reported on the exceptional but disappointing events of 1886, which supposedly gathered around fifteen fasters in Europe and the Americas who practised 'a new genre of sport' and struggled to resist the absence of food: 'In 1886, Paris witnessed the birth of a new type of sport. For a few months, the news was full of stomach-strength tricks: there were about fifteen individuals in Europe and America who were fasting, preparing to fast, or who had just gone twenty, thirty, forty, fifty days without eating.'[56]

But, above all, as described by some of his Italian biographers, Succi was a wonder of the nineteenth century, with impressive, synchronic press coverage that crossed continents.[57] In May 1886, for instance, the

Notre Dame Scholastic, the student magazine of the University of Notre Dame (Indiana), reported on Succi's recent achievements in Europe.[58] Equally, in 1888, an apparently local, peripheral event such as the thirty-day public fast by Succi in Barcelona was given a striking amount of coverage in the press of the Antipodes. The *Poverty Bay Herald*, a daily newspaper in New Zealand, described the end of this public performance in the following terms:

Succi, the Italian faster, has now ended his thirty-days fast [in] Barcelona. As on previous occasions, Succi's first meal after his long fast was decidedly heavy. He ate abundantly of sardines in oil, butter, tapioca soup, fried brains, filet de boeuf, spinach, steak, roast chicken, fruit, and sweets. He also drank Chianti, soda water, and Malaga. This meal will give some idea of the strength of the Italian faster's stomach, after which, moreover, he smoked some green Havana cigars.[59] Succi has lost a good deal in weight during his fast, but his spirits have been throughout excellent. He took plenty of exercise, particularly in fencing, in which he is an adept. The narrow watch kept over him during his fast forbids the suspicion that this was other than genuine. I hear that the conclusions of the Spanish doctors are that Signor Succi's so-called 'elixir' is really of value, and may be of great benefit to science in exceptional circumstances. At the same time, it is the opinion of Spanish medical men who have carefully watched Succi during his thirty days' fast that he is an exceptionally constituted person, and could live for considerable periods on the nourishment stored in his own tissues.[60]

In 1890, the *Victoria Daily Colonist*, a Canadian newspaper from British Columbia, described Succi's fast in New York in the following terms:

'Fastest Man in the World. Succi, the Successful Abstainer, Arrives to Challenge Dr Tanner'. New York, Sept. 10. Giovanni Succi, the Italian faster, arrived from London Saturday on a mission to knock out Dr Tanner's record and his own by fasting forty days with him. He has a trunk full of documents to prove that he is the real Succi. These consist of volumes bound in red and black Morocco, with the signatures of 200 physicians who watched his fast; the record of the observations of Paris physicians, and scrap books filled with accounts of his accomplishments in Italian, Spanish, French and English. Succi fasted thirty days in Lisbon, thirty-five in Brussels, and forty in London. He has gold medals from the Paris exposition and the London exhibition, with his profile of bas-relief. The faster is a man of medium size. His flesh is as hard as iron, while he is of an extremely nervous organization. The force that sustains him, he says, is spiritual, and after a few days of fasting he can take up foils and vanquish the best fencers. A committee is to be appointed to watch him fast.[61]

Both news items reflected the global dimension of these fasts. Whereas the *Poverty Bay Herald* stressed the details of a fasting perfomance, a sort of ceremony which probably interested global readers and was usually reproduced in a more or less standardised narrative (see Chapter 2), the *Victorian Daily Colonist* pointed out the multiple proofs of the faster's reliability, the legitimation of the medical class, and the international exhibition juries of the controversial practice.

Playing with the readers' potential attraction to sensationalism, the daily press often placed hunger artists on the fringes of fraud and charlatanry. In 1896, in Vienna, for example, suspicion of Succi's potential fraud very quickly spread through the media. An American medical treatise soon reflected on the Austrian crisis in the following terms: 'It has come to light in his latest attempt to go for fifty days without food that he privately regaled himself on soup, beefsteak, chocolate, and eggs. It was also discovered that one of the "committee", who were supposed to watch and see that the experiment was conducted in a bona fide manner, "stood in" with the faster and helped him deceive the others. The result of the Vienna experiment is bound to cast suspicion on all previous fasting accomplishments of Signor Succi, if not upon those of his predecessors.'[62] In 1909, two simultaneous fasts took place in England and Germany. In the English case, a doctor named M. Penny fasted for thirty days for the sake of the experimental study of the therapeutic consequences – à la Tanner – of that period of abstinence on his own health.[63] At the same time, in Berlin, the medical student Claire du Serval (see Chapter 4) self-experimented with her own body for periods of six to ten days of fasting, only drinking pure water as a therapy to cure several illnesses. She recommended that young, elderly, and ill people fast for two days per week as a useful therapeutic strategy.[64]

Obviously, readers in New Zealand, Canada, Vienna, and Paris did not have the sense of belonging to a single community, but in practice they shared sound data and reacted to global events that in many cases included hunger artists' performances. Overcoming long distances, news on hunger artists, resulting from a general expansion of the communications networks, contributed to a new sense of synchronicity. In 1883, a memorandum sent by Reuters from London to all its agents and correspondents can help us to grasp the context in which news about hunger artists appeared in the 1880s:

> In consequence of the increasing attention paid by the London and English provincial press to disasters of all kinds, agents and correspondents are asked to be good enough in future, to notice for London all occurrences of the sort. The following are the events which should be comprised in the service: Fires, explosions, inundations, railway accidents, destructive storms, earthquakes, shipwrecks attending with loss of lives, accidents to British and American war vessels and to mail steamers, street riots of a grave character, disturbances arising from strikes, duels between, and suicides of, persons of note, social or political, and murders of a sensational or atrocious character ... London, November 1883.[65]

Although they were not explicitly mentioned in this Reuters note, hunger artists appeared as persons of social note, worth being reported in the public sphere of the late nineteenth-century global press.

Networks

Medical and commercial relations played an active role in the construction of the land of the hunger artists, as a fluid, network-like system of global interest on human resistance to inanition. From Boston to St Petersburg, from Berlin to Tokyo, doctors were part of the geography of late nineteenth-century physiology, a discipline that had also acquired an important global dimension. Coming from different traditions and perceptions of the functioning of the human body, and often dealing with disagreements and controversies, those physicians exchanged experiments and utilised hunger artists as objects of scientific inquiry. In the second half of the nineteenth century, experimental physiology grew through an increasing number of research schools with regular interest in hunger artists (see Chapter 3). German research schools became salient nodes of the network and key references for training. Johannes Müller[66] and later Nathan Zuntz in Berlin,[67] Carl von Voit in Munich, and Carl Ludwig in Leipzig were some of the leading names. Following Liebig's tradition in Giessen in animal and vegetable chemistry,[68] German laboratories provided an unparalleled richness in teaching – *Lehrenfreiheit* with itinerant lectures – and active research with the involvement of students in the experiments.[69] French physiologists such as François Magendie, Claude Bernard, and, later, Charles Richet also had their say, often not sharing radical materialistic approaches with their German colleagues, but contributing to a geographical space of circulation of persons and experiments. In the Anglo-American world, two milestones of the network were the foundation of the Physiological Society in London in 1876 and Michael Foster's (1836–1907) edition of the new *Journal of Physiology* in 1878. Foster had a place in the foundation of the Cambridge biological school, and his physiology textbook had a great impact on the establishment and spread of the discipline.[70] On the other side of the Atlantic, the American Physiological Society was founded in 1877. There, a pupil of Ludwig's, Henry Bowditch (1840–1911), played a leading role, later becoming dean of Harvard Medical School. Bowditch's training tour in Europe had also included Claude Bernard's in Paris, so he acquired great experience of the different physiological schools and their research priorities.[71] It was in this dense map of physiology schools that hunger artists found their place.

Public fasts often became focuses of attraction for the medical community. In 1886, for example, at the Grand Hotel de Paris, Merlatti welcomed distinguished visitors such as Dr Thomas Linn, a former member of the medical board of Tanner's fast and correspondent of the *New York Medical Journal*, who had specifically come to Paris to report on Merlatti's medical condition during the fast, and to compare him with his American predecessor.[72] Merlatti also drew the attention of other relevant names such as Dr Oscar Jennings, the correspondent for *The Lancet* in London, French local physicians such as Dr Laisant and Dr Bonnafont from the Paris Academy of Medicine, and other professors of the École de Medicine.[73] Luigi Luciani (1840–1919), who studied Succi's fast in Florence in 1888, was very well placed in the international network of the contemporary schools of physiology (see Chapter 3). After studying medicine in Bologna, Luciani worked with Ludwig at the internationally renowned Institute of Physiology, in Leipzig, a key node in the network. Luciani's career later followed a particular geography within the confines of Italy, but was full of international connections. In 1893, he was appointed to the chair of human physiology at the University of Rome, where he headed the Istituto Fisiologico and later became Rector in 1898. In Rome, he succeeded the famous physiologist Jacob Moleschott (1822–1893), who had focused much of his research on nutrition.[74] Luciani led a highly renowned school of physiology, in which the physiology of fasting had an important place.[75] Among his pupils, it is worth mentioning Giulio Fano (1856–1930). Having trained with Moleschott and Ludwig, Fano became Luciani's assistant in Florence and director of Laboratorio di Fisologia. Like Luciani, Fano disapproved of the division of physiology into different specialities.[76]

Luciani's Italian network and its international connections provided a cosmopolitan view of the study of hunger artists, but local varieties and plural appropriations were inevitable. We can recall the Parisian Academy of Medicine's lack of interest in public fasting performances during Succi and Merlatti's 1886 contest, which contrasted with the enormous interest in that performance on the part of numerous doctors. Likewise, Succi's fast at the Royal Aquarium in London in 1890 did not garner much interest among the local medical class, who mainly considered that kind of charlatan as useless for medical research (see Chapter 2).[77] In other places, however, such as in the case of Barcelona in 1888, some local doctors perceived Succi as offering an opportunity to reinforce their medical views (see Chapter 5). In addition, debates on fraud during fasts and on the role of official medicine as fighting against the risks of charlatanry also suffered from tensions between the local and the global. As discussed earlier, while in 1896 in Vienna the medical board was questioned, in other places its reputation remained intact. Nevertheless, beyond national boundaries and local interests, such disputes and disagreements strengthened the ties of the international network. In the land of the hunger artists, medical controversies obviously did not exclude teamwork or scientific collaboration.[78]

In 1907, the recently appointed director of the Carnegie Nutrition Laboratory in Boston, Dr Francis Gano Benedict, began his tours of European labs – which lasted until the 1930s – focusing on their study of metabolism and nutrition physiology.[79] He represented a generation of American physiologists who had been trained in Germany, and who deeply appreciated international collaborations. In his seven tours of Europe he exchanged ideas on metabolism and the physiology of fasting with many research groups and shared observations on the material culture of the laboratories.[80] Benedict was particularly interested in human metabolism from birth to death, diabetes, the effects of a reduced diet, fasting, and racial differences, so hunger artists were part of his scientific agenda to be shared with his European colleagues. Benedict aimed to acquire enough experience and foreign references for the construction of an ideal laboratory for the new branch of nutritional physiology, but he also praised the value of tacit knowledge involved in the experiments on metabolism and respiration as well as the importance of establishing an equal, peer-to-peer collaboration with his colleagues (see Chapter 3). Benedict visited university laboratories, vocational schools, medical clinics, experimental agricultural stations, and research centres in Britain, continental Europe, and Russia.[81]

In 1915, in his *Study of Prolonged Fasting*, Benedict described the international efforts made in the realm of experimental physiology to study the processes of human inanition and the use of hunger artists as objects of scientific experimentation.[82] He referred to a very rich network of medical authorities on fasting, who had introduced him to the field: Luigi Luciani (at that time in Rome), Giulio Fano (Florence), Nathan Zuntz (Berlin), Franz Tangl (Budapest), Robert Tigerstedt (Helsinki), and Graham Lusk (New York). He also mentioned several collaborative endeavours for the study of particular fasts in the context of the physiology laboratories and the shared results in the network: Hermann Senator, Zuntz, Lehmann's study on Cetti and Breithaupt (1887), Luciani's study on Succi (1888), Noel Paton and Ralph Stockman on Jacques (1888), Drs Hoogenhuyze and Verploegh on professional fasting woman Flora Tosca (1905), Theodor Brugsch's study on fasting woman Auguste Victoria Schenk (1906), and Edward P. Cathcart on Beauté (1907) (see Chapter 3). Benedict stressed in particular that the network of physiologists had worked for years on the case of Succi, from Milan in 1886 to Hamburg in 1904.[83]

Equally, other experiments on the physiology of fasting took place, for example, in Russia.[84] This was the case of the Imperial Military Medical Academy in St Petersburg, under the direction of Professors Viktor Pashutin, Vasily Danilevsky, Ivan Pavlov, and A. A. Likhachev.[85] In Tokyo, at the Aoyama Clinic, hunger artist Kozawa was studied by physiologists R. Watanabe and R. Sassa, and the results of the study were shared in textbooks.[86] As discussed in detail in Chapter 3, the quantification of the human metabolism relied on the difficult standardisation of a key instrument, the respiration calorimeter. Benedict had inherited it from the tradition of his masters – from Carl von Voit's school in Munich to Wilbur Olin Atwater in the Wesleyan laboratories in Middletown, Connecticut, on the other side of the Atlantic – but wanted to transform it into a real universal respiration apparatus, a very demanding scientific challenge at the time.[87] It was in this kind of collaborative enterprise among different schools of physiology that hunger artists were key players mediating among different styles of scientific expertise.

But the network also had a commercial side, which is worth appending here. Impresarios played a significant role in it, not just for business, but also delineating some rules of the game – often with the supervision of the medical boards – through strict conditions in written and signed contracts. In 1886, for example, Succi was introduced to scientific circles by Dr Barberi-Borghini, but the business was in the hands of an impresario named Lamperti, who established – not without disagreements – the pattern of the public performance.[88] In 1888, in Madrid, Succi made a deal with the local impresario Felipe Ducazcal, the owner of Teatro Felipe, which hosted his fast.[89] One of Succi's impresarios, Achille Ricci, accompanied him on some of his trips and itinerant fasts, especially in North and South America.[90]

In 1890, in New York, at the end of Succi's fast, the famous American traveller and eccentric entrepreneur George Francis Train (1829–1904), known as 'Psycho Train', praised that great achievement. The press stressed that he 'outdid himself in fervid oratory and general pyrotechnics, and brought the great fast to an end in a blaze of glory for all concerned'.[91] Psycho Train's heterodox culture (he was from a Methodist background), his adventures, and his vegetarianism made him feel close to Succi's transatlantic endeavour, and he expressed deep sympathy for Succi's cause. But there was more to it than that. In 1878, Psycho himself had attempted to fast for some days after adopting 'the novel theory that eating is an acquired habit, and that it would be an excellent thing to abandon it'.[92] He was in fact one of Succi's many rival fasters; Psycho particularly welcomed Succi's visit as part of his strategy to raise the interest of the public. In the city, there were even rumours that the two men had planned a 'great hundred days' fast', which never actually took place.[93] Psycho Train was another of the varied characters who, beyond the strict medical community and the supervision by the medical boards during the fasts, accompanied hunger artists on their journeys, tied new personal relations to strengthen the network that built the 'land', and provided the whole endeavour with a sense of cosmopolitanism, sensationalism, and controversy. Likewise, Humberto Fiandra, another of Succi's impresarios, also described in the press a fast in Turin in 1898, after a successful tour across Italy. Praising Succi for his return to Turin after years abroad, Fiandra organised a long fast at the city's Odéon theatre.[94]

***

The land of the hunger artists had its genuine geography. Often taking place simultaneously and synchronically reported, the numerous public fasts marked the landmarks of a geography of travels that resembled navigation itineraries in often uncharted oceans. As Succi's adventures in Africa epitomised, travels became a mythical reference for the legitimation of the metier, but at the same time an intrinsic part of the itinerant nature of professional fasters, an indispensable section in their curriculum, when they sought scientific recognition and profitable business in every local context. The arrival of hunger artists in a new city with a suitcase full of diplomas and certificates from other places reinforced their prestige and attracted new customers, to the satisfaction of their impresarios. Although disagreement and suspicion about this kind of spectacle never disappeared, daily reports in the press of the vital signs measured by medical boards in any corner of the globe made hunger artists widely known, and news of their performances was shared (and tacitly legitimised) by millions of readers.

The itineraries of professional fasters fitted in the general pattern of the itinerant nature of charlatans, but the nature of the land of our fasters seems subtler and more complex than a simple travelling existence selling miraculous cures. Their impressive presence in the press, the deep involvement of the international network of physiologists in the scientific study of prolonged abstinence, the commodity-like nature of the metier, and the role played by impresarios and medical boards in the establishment of specific rules to follow during fasts in every local context: all make the case of the hunger artists particular. In addition, the never-ending tension between local specificities and global patterns of their behaviour travelling from one side of the planet to the other paradoxically reinforced their network-like nature. Just as controversies, natural catastrophes, and social scandals became useful topics for a growing sensationalism, issues such as charlatanry, fraud, human resistance, and human suffering and debates about the limits of the human body put hunger artists on the map. However, the specific places in which the fast took place in every local context require another level of geographical analysis, another spatial turn, which I shall explore in the next chapter.

2 Performances

As Kafka described in his short story 'Hungerkünstler', hunger artists' loneliness and marginality could be understood as a sort of metaphor for the never-ending challenge for the artist (the writer, Kafka himself) seeking public appeal and recognition. As Maud Ellman discussed some years ago, it was the public gaze, the active role of the audience, as a reaction to the performance that became a driving force for hunger artists to fight against inanition and prove their resistance in reaching the risky territory of starvation.[1] This chapter therefore attempts to dissect the relationship between hunger artists and their audiences, as well as the way in which that ritualised relationship was embodied in particular places that shaped the epistemology of the performances.

Place matters when analysing dynamic processes of knowledge construction, so the locations of public fasting performances have enormous potential in analytical terms.[2] On numerous occasions, the disciplined atmosphere of a physiology laboratory was combined with fasters fencing, running, walking, swimming, or climbing towers and cathedrals. In all these sites, hunger artists constructed their own identity and interacted with the varied audiences. Aileen Fyfe and Bernard Lightman described some years ago, in their *Science in the Marketplace*, how nineteenth-century science, particularly in Victorian Britain, appeared in countless locations: panoramic shows, exhibitions, galleries, city museums, country houses, popular lectures, and domestic conversations, which shaped the way in which knowledge circulated and was made by the different actors.[3] In the case of hunger artists, their dialogue with doctors, but also with journalists, curious onlookers and visitors, and the general public in particular places, shaped contemporary knowledge of fasting, inanition, and starvation and influenced doctors' decisions and strategies.[4]

Reconstructing the spatiality of the rituals of public fasts can help us to understand how a definite space constrained this kind of performance. This is therefore another level of hunger artists' geographical analysis (see Chapter 1), in which objects and places matter. Beyond a simple topography of cities and countries, in which professional fasters performed, this chapter brings to the fore a huge range of unexpected places, which linked the artists with their audiences, far from the traditional spaces of scientific practice. The choice of venue, the advertising campaigns, the design of the cage or the sealed room in which to spend days in inanition, the day and night supervision, the scientific supervision by medical committees, the regular visits, the negotiations with the impresario, the faster's public statements, the last meal before the fast, and the celebration banquet at the end of the days in inanition: all constituted a situated protocol with slight variations in each local context, with many commonalities, which might explain the hunger artists' complex journey between science and spectacle.[5]

To close the circle, which had started with hunger artists' voracious feeding on the eve of their performance, a cheerful audience used to celebrate the completion of the fast with an opulent banquet. Kafka reproduced in fictional terms his own experience as a witness of that pompous public ritual with the following words: 'on the fortieth day the flower-bedecked cage was opened, enthusiastic spectators filled the hall, a military band played, two doctors entered the cage to measure the results of the fast, which were announced through a megaphone, and finally two young ladies appeared, blissful at having been selected for the honour, to help the hunger artist down the few steps leading to a small table on which was spread a carefully chosen invalid repast'.[6] This was a general pattern. On 26 October 1886, in Paris, on the eve of the beginning of his fast, Stefano Merlatti ingested a full fatty goose including the bones, a 1 kg beef fillet, two dozen walnuts, and some vegetables.[7] In 1888, at the end of his fast in Barcelona, and before all the members of the supervision committees, the press, the overseers, the jury of the Exhibition and the general public, Succi washed out his stomach for the last time, did some fencing, and drank his first mug of soup. Then, the jury awarded him a gold medal in recognition of his heroic stay at the International Exhibition's Science Pavilion. Likewise, in 1905, in Mexico City, the end of Succi's fast was celebrated with a generous meal, offered in this case by the restaurant Maison Dorée.[8]

These pompous celebrations were the last step in a set of standardised performances, which normally began with a detailed description of the site in which the fast would take place; the establishment of medical supervision committees; a contract with an impresario; and an agreement with local scientific institutions and the municipality. It also included a well-defined set of urban, public performances (riding, walking, running, fencing) and the scientific experiments that accompanied them; visits by the public at large; and wide coverage in the press by journalists. Sometimes, hunger artists also used interpreters as key mediators to 'translate' to the public what was supposedly happening in the inner rooms or cages in which the fast took place. On other occasions, they hired a lecturer who introduced them to the local audience and accompanied the hunger artists from city to city.[9]

The case of Succi's fast at the Royal Aquarium London, compared to other key places such as the Sheffield Jungle, reveals the intimate coexistence of hunger artists with freak exhibitions and oddities. Although they lacked the spectacularism of other kinds of human deformities and freak shows, their 'monstrosity' – their capacity to fast for long periods of time with scarcely any deterioration of their physical condition, but clear signs of emaciation of their body – made them, for quite a long time, appealing objects of curiosity and attracted a refined bourgeoisie – including many women – when exhibited in theatres, world's fairs, noble apartments, and hotels. Nevertheless, when hunger artists descended to the hell of freak shows, they also captured the attention of the working class. In both cases and in all venues, they became a lucrative commodity. Public fasting was a spectacle of an enormous richness, in which the noise of the masses, in a broad sense – from medical committees to journalists, popularisers, and the general public – made doctors' attempts at 'isolating' the faster, for supposedly 'objective' scientific study, an impossible endeavour. However, this was an intrinsic part of the game, in a show that had to continue regardless of the circumstances. Doctors' unease in the face of the unlimited opinions about the art of fasting was inevitably counterbalanced by their ambition to obtain social recognition and professional respect. As in many other contemporary processes of knowledge in transit, tension between expert and lay views of the same event was inevitable.

After establishing the location for 'putting hunger artists in their place' – to follow David Livingstone's spatial approach to the geography of science[10] – this chapter discusses how different places blurred the boundaries between professional fasters' performances and contemporary freak shows and oddities, and how the commotion of the audiences apparently fogged but, in the end, stimulated the artists' creativity and their overseers' authority. As literature scholar Breon Mitchell rightly pointed out in his study of Kafka's short story, in 1890, at the peak of their popularity, hunger artists became an appealing spectacle for urban bourgeois crowds, and daily news of them spread in newspapers around the world. However, it was not obvious where the spectacle was, nor how specific sites shaped the epistemological consequences of the performances.[11] This is a crucial challenge that the sections that follow will explore carefully.

Places

Hunger artists' performances were usually sealed off from the public in restricted spaces. A public fast always took place at a site in which the artist stayed confined for a long time under the supervision of the committees and the gaze of the spectators, but all of this was combined with frequent visits and outings through the city. The open–closed ambiguity of the fast acted as a powerful method to attract public recognition and customers' interest. In fact, because of the intimacy of the sealed cages, the comfort of the rooms to welcome visitors, and the occupation of the urban space, the spatiality of hunger artists' performances took on different shapes in every local context, but also had common features which are worth exploring here.

In 1886, for example, in Milan, Succi went for a swim in the Ticino river and returned on foot to the venue, where 600 people watched him in a fencing match.[12] The same year, the impresario Mr Lamperti exhibited Succi in an apartment on rue Peletier, in Paris. Entrance tickets cost two francs during the week and one franc on Sundays. Since Succi did not attract many visitors there, Lamperti decided to move him to theatres such as the Olympia and Eden, but also to rollercoasters, a very popular mechanical amusement device at that time.[13] In his contest with Succi, Merlatti planned a journey from the Grand Hotel to the Bois de Boulogne on the seventeenth day of his fast.[14] The next day, the daily reports noted that he continued painting and drawing, receiving visitors from noon to 6.00 pm. At 6.30 pm he went for a short walk until 8.00 pm, always accompanied and supervised.[15] In 1888, Succi stayed at the Accademia Medico Fisica in Florence for medical experimentation, but at the same time filled the Teatro Imperiale with an upper-class audience of 2,400 people.[16] In 1890, in New York, Succi crossed the Brooklyn Bridge in a popular parade. He also appeared at the 1891 Chicago World's Fair, a Boston dime museum (1891), Marseille's Palais de Cristal (1895), the Hotel Royal at the Prater in Vienna (1896), the Amfiteatro romano in Verona (1897), and the VII Genoa Industrial Exhibition (1901); he was sealed in a box in a cinema in Bologna (1907).[17] In 1897, the site of Succi's fast in Verona was described as follows: 'Succi in a box: a bed … a thermometer, a pair of scissors, a razor, a corkscrew, a piece of soap, two pencils, a notebook, a pack of candles, 7 bottles of Vichy water, and a small bottle of "that famous elixir that he invented and that did not make a fortune", a good French novel: Zola's "Pot-Bouille", 50 centimes for admission, gymnastics, fencing, horse riding.'[18]

Following a similar pattern, in 1888, in Barcelona, outside the restricted area of the International Exhibition, Succi also sought to prove

Figure 2.1. Giovanni Succi on horseback in Barcelona on the twenty-seventh day of his fast, with some members of the supervision committee, in front of the Arc de Triomf, at the entrance of the 1888 International Exhibition.

the authenticity of his performance in other places. Close to the end of his fast, the city as a whole became a jury for his achievements. Accompanied by several members of the medical committee, Succi walked for an hour and a half through the streets and rode a horse to show his unaltered physical skills and strength (Figure 2.1). Two days later, he practised fencing with his teachers.[19] Gymnastics, horse riding, and swordplay became effective public performances of his healthy state.[20] By purchasing a ticket for two reales (half a peseta), the general public could visit Succi in his room at the Palacio de Ciencias.[21] In the company of medical doctors and journalists, Succi rose up to the height of 300 metres in one of the greatest attractions of the Exhibition, the tethered balloon (*globo cautivo*). No dizziness or any kind of sickness or distress was reported. The balloon was installed inside the exhibition area, and provided a magnificent view of the whole city. As advertised in the press, the balloon went up daily from early morning to night. It was designed to attract as many visitors as possible, to complement other amusements and provide incomparable views of the city from the air.[22]

In Madrid, also in 1888, the medical committee was fully responsible for the details of the experiment. It decided the place of the performance, and the public were not allowed to have physical contact with the faster. All entrances to the salon in which the faster was exhibited would be locked and under medical supervision. The committee would have a separate room, isolated from the public, for daily discussions, meetings, and tests on the faster's samples. From midnight to 8.00 am only the medical committee would have access to Succi.[23] Some years later, in 1905, this time under the auspices of the impresario Manuel Castro, Succi arrived in Mexico City from his former stay in Havana, where he was also subjected to detailed medical supervision at the Teatro Albisu.[24] He planned to fast at the Teatro Orrin, but he finally, on 25 April that year, ended up in a box at the Teatro Principal to stay there for a month without food,[25] under the supervision of doctors, journalists, medical students, and representatives of scientific societies of the capital and members of the Consejo Superior de Salubridad.[26] On 13 May, at the end of his fast, nobody accepted a fencing contest with him, since competing against a man who was close to starvation after thirty days without food did not seem honourable.[27]

In other cases, sealed cages and supervision committees conquered the city in very popular urban parades. In 1889, during the Paris International Exhibition, and after being seen on the rollercoaster, Succi climbed the Eiffel Tower, accompanied by members of the medical committee.[28] A year later, in New York, the *New York Herald* and Succi's observers organised a Brooklyn parade, on the fifteenth day of his fast, with six horses with riders, two coaches, a regiment, and a band that accompanied him from his fasting place on 23rd Avenue to the Brooklyn Bridge and back again, dismounting at the *Herald* offices, and moving upstairs and downstairs very quickly.[29] In 1892, Succi hiked up Vesuvius. Also, in 1890, in London, asked by an anonymous journalist about his plans regarding the place where he planned to do his performance, Succi answered:

> so far, nothing has been settled. Last year, when fasting in Paris, I went up the Eiffel Tower on the fifteenth day of my fast. [In] Rouen, I ascended the cathedral spire on the twelfth day of my fast; and [in] Brussels only lately I astonished the committee by running up to the top of the Colonne du Congrès in two minutes and a half. It is in this taking of exercise that I differ from other fasting men, like Tanner and Merlatti. During their periods of fasting they remain supine, inactive, but I can fence and walk, play billiards, climb, and swim; and at the end of my fast I am as strong as at the beginning.[30]

Sealed cages and open urban itineraries also intermingled in other cities. There are some references to his tour across the United States and Canada[31] in which Succi's popularity seemed to rival that of Stanley, his former travel companion in Africa.[32] The press also referred to Tanner's attempt to meet Succi in person at the Chicago World's Fair venue.[33]

Later came other performances in local industrial exhibitions such as Genoa, in 1901, where he appeared sealed in a cage for the strict supervision of his fast.[34] In 1896, Succi fasted in Vienna, at the Hotel Royal, where his performance was extensively reported in the *Illustrirtes Wiener Extrablatt*.[35] The pattern was again similar: with a medical committee supervising him, a considerable crowd waiting for him, and even a photographer, which resulted in the commercialisation of signed portraits of Succi as a celebrity. In the entrance hall, Succi wore knight's armour as a symbol of his isolation and willpower. Over Easter, he welcomed more than 500 visitors, including many women who offered him flowers, and Archduke Leopold, with whom Succi shared the motivations behind his impressive resistance to hunger and inanition.[36] Some years later, in 1905, the same city hosted a German hunger artist, Wilhelm Bode, with an Italian artistic name, Riccardo Sacco, who fasted for twenty-one days in a café at the Prater, again confined in a small cabinet with well-sealed glass sides.[37] A year earlier, in 1904, the press described Sacco's controversial fast in Berlin, in which: 'he had himself bricked in and could be seen only through a small grated window. But students conceived the idea of Sacco as a fraud and stoned him through the window till the police interfered, arresting some of his persecutors and digging Sacco out and forbidding any further public exhibition of starvation.'[38] In fact, the crowd's excitement by the suspicion of a potential fraud became frequent in fasting rituals. In 1904, Sacco had also fasted at Oktoberfest in Munich.[39]

Regarding the power of place, a reconstruction of Succi's ritual of public fasting in 1890 at the Royal Aquarium London is particularly enlightening. Succi was confined day and night to a poorly ventilated, gas-lit apartment, with volunteer physicians and medical students from the London hospitals. However, the press expressed doubts on the reliability of the experiment.[40] Controversies apart, though, Succi's fast was lavishly depicted in the *Illustrated Police News* (Figure 2.2). In the image, the faster appeared at the top as an adventurer (right) and sportsman (left), from his mythical trips to Africa to his fencing skills. At the bottom, the actual place in which he fasted was carefully depicted: a room with a bed, a table, and a couple of chairs where Succi sat, read, wrote, and talked, next to three overseers with laboratory equipment for the daily analyses and medical checks.

Finally, and closer to the public, in the bottom right corner of the illustration, an interpreter appeared as a conductor of the whole performance, who potentially interacted with the visitors and translated Succi's words into English: 'an attendant showman discourses to the crowd of visitors'.[41] On the interpreter's left, a small table holds 'autographs to

Figure 2.2. 'Succi, the Fasting Man'. Giovanni Succi's activities and surroundings at the Royal Aquarium London.

sell with copies of his portrait'.[42] It is worth quoting here the detailed description of Succi's fast at the Aquarium that appeared in the *Illustrated London News*:

The sequestered and canopied space in the Royal Aquarium at Westminster, prepared for his abode, has been daily visited by numbers of people, willing to pay an extra fee, and their curiosity has been gratified by seeing an Italian gentleman, of sallow complexion and evidently, though muscular, without an ounce of spare flesh, neatly dressed and seeming quite at his ease, sitting at a small table raised upon a platform, behind which is the bed where he sleeps at night. At a lower table sit two or three gentlemen supposed to look after their business, with whom may be associated some volunteer belonging to the medical profession – if not a qualified practitioner, it may be a student of the medical school of one of the London hospitals; but we are unable to state what official authority, from the recognised staff of any of those institutions, has made itself in any degree responsible for observing a case of such great scientific interest.[43]

Another image showed Succi welcoming visitors – including ladies who, as in other cases, queued to see him in person and to receive a signed picture.[44] Succi also appeared smoking, riding a horse, showing his emaciated face on the thirty-ninth day of abstinence from food, and eating a frugal soup at the end of his fast at a table which contained bottles, probably of his famous liquor and the mineral water he drank (see Chapter 5). At the back of the stage, members of the medical committee bring more food to him as part of the celebration banquet that was common at the end of the fast. In fact, the whole performance at the Royal Aquarium is a good representation of the open–closed spatial nature of public fasting, which included a huge range of actors, objects, and sites, all of them relevant at different levels in epistemological terms (Figure 2.3).

The obsession with sealing the faster away from public 'contamination' usually ended up, paradoxically, with the public involvement of other actors who influenced the way in which these fasts were perceived in their time. Hunger artist Pappus, of South American origin, was known as the 'Man in a Bottle'.[45] In 1905, in the Crystal Palace Theatre of Varieties at Leipzig, Pappus's daring spectacle resonated even in the United States as the *Washington Post* reported in the following terms:

At first, he lay for a week in a glass case, like the snow maiden in the fairy tale, then he existed on compressed air under water in a peculiarly constructed submarine boat, until at last he came to his present form of eight-day incarceration in the original six-sided, bottle-shaped glass house into which he stepped in Leipzig. This receptacle allows only the smallest movement on the part of the hunger artist inside the glass cage. Pappus has no comforts to mitigate the peculiar character of his experiment. He presents a mummified appearance, standing

Figure 2.3. 'The Forty Days' Fast at the Royal Aquarium'. Giovanni Succi's activities and visitors during his fast at the Royal Aquarium London: 'The congratulations of friends', 'Succi enjoys a smoke', 'Succi on the 39th day of his fast', and 'Signor Succi takes horseback exercise'.

all day in the narrow glass space, in which he can make only the slightest turning movement. At night, a horizontal position is allowed. The demonstration of this new hunger man differs considerably from that of the Italians, Merlatti and Succi, and Dr Henry Tanner, the American, who retained their liberty.[46]

In fact, the sealed spaces acted as a symbol of control, scientific supervision, and fraud prevention. They covered a huge range of options, from Pappus's thin glass bottle – not far from the respiration calorimeters that contained hunger artists in experimental laboratories (see Chapter 3) – to more spacious rooms in hotels that could host a large number of visitors, spaces to welcome the medical committee, and other sites to vividly interact with crowded audiences, as in the Royal Aquarium. But the sealed, supervised areas always had their counterbalance in more open spaces: from World's Fair pavilions to organised city tours, fencing contests, and the spectacle of climbing monuments. In 1890, interviewed by the *Pall Mall Gazette*, Succi prided himself on his physical strength in former fasts, not only to exercise, ride horses, run, walk, and fence, but also to conquer public urban spaces as part of the marketing of his spectacle.[47]

Places, therefore, had a relevant epistemological status in hunger artists' performances. They acted as mediators between fasters, doctors, impresarios, supervisors, and the varied audiences, and contributed to blurring the boundaries between science and hype and to a fluid co-production of knowledge on human inanition. As Livingstone highlighted in his *Putting Science in Its Place*, the location of a scientific endeavour – and, why not?, a performance, a practice or a ritual – can make a difference to the conduct of knowledge and even affect its content.[48] In that context, the rich spatial nature of hunger artists' performances contributed to the development of fluid, often controversial knowledge about human inanition and made experts' attempts to 'isolate' these cases inside the academic realm truly untenable.

Freaks

In that fluid spatiality, the boundaries between freaks, charlatans, and hunger artists were subtly undefined. In 1905, the press reported, for example, that: 'Herr Pappus calls himself a hunger artist and seems to regard his curious profession with great respect',[49] but Pappus used to introduce himself as a fakir and faster. Although it was often presented in public as pursuing an educational or scientific value, in practice, the exhibition of people with 'alleged and real physical, mental, or behavioural anomalies' pursued amusement and profit, and became an intrinsic part of the commodification trends of the second half of the nineteenth century.[50]

A huge range of human oddities – obese people, giants, dwarves, living skeletons, armless and legless wonders, people with albinism, Siamese (conjoined) twins, fasting girls, and hunger artists – filled popular spectacles and urban recreation. Places such as the Royal Aquarium in London, the American dime museums such as those in Boston and New York (which proliferated across the country precisely in the period of 1880–90, at the peak of hunger artists' success), the Sheffield Jungle, and the Crystal Palace Theatre of Varieties in Leipzig, among many other venues, welcomed different sorts of oddities. In competition with circuses and other amusements, dime museums, in particular, became natural sites for the shabbiest oddities and promoted 'fraud, misrepresentation, and exaggeration'.[51] The first type of freaks belonged to what Robert Bogdan called 'the exploration of the non-Western world' as exotic creatures of colonial origin who were exhibited in Western urban contexts. The second consisted of 'monsters', which raised particular interest among physicians and contributed to the science of teratology. In the late nineteenth century, though, both types merged into a general category of 'freaks'.[52] In that 'monstrous' universe, there was indeed a place for hunger artists too.

In 1890, in spite of the apparent satisfaction of the witnesses of Succi's fast at the Royal Aquarium, the institution's reputation posed problems for many doctors to certify the fast as a performance, since many perceived it as another freak show, as a spectacle of oddities in a low-class atmosphere.[53] With some exceptions, medical journals commented on the event in a sceptical way and often reflected on the useless nature of such experiments for science. Paradoxically, though, Succi's prolonged fast had great visibility. Although originally conceived in the 1870s as an ambitious entertainment venue including an aquarium, a concert hall, a theatre, a picture gallery, and a skating rink, among other facilities, and enjoying a priori the support of the Duke and Duchess of Edinburgh, the Royal Aquarium London progressively became a 'piece of humbug', as described by the *New York Times* in 1876.[54] The morality of the clientele also degenerated, often attracting thieves, drunks, and prostitutes. In the 1880s, with no fish or water, the Aquarium ended up as a freak show venue to hold spectacles,[55] in an atmosphere that the press characterised as deplorable for scientists. As a history of the lost London theatres described some decades later:

> The Royal Aquarium ... was intended to be a sort of Crystal Palace in London within easy reach of Charing Cross, a covered-in promenade for the wet weather, with the glass cases of live fish thrown in. In truth, the attractions of the place soon began to be very 'fishy' indeed. Ladies promenaded there up and down without the escort of any gentleman friend (till, maybe, they found one) and the appeal of the management to sensation lovers was very wide indeed. Bare-backed ladies dived from the roof or were shot out of a cannon, or sat in a cage covered with hair and calling themselves 'Missing Links'. Zulus, Gorillas, *Fasting Humans*, Boxing Humans and Boxing Kangaroos, succeeded one another in rapid changes, and failed in time to attract.[56]

During Succi's fast at the Royal Aquarium 1890 (Figure 2.4), and in spite of that freak show atmosphere and frequent reluctance in the press, Dr G. N. Robbins offered his name and professional credentials to supervise the experiment, and Drs William Powell, A. Burrow Burt, and Alfred E. Pike, from the Westminster Hospital, were in charge of Succi's medical supervision and the chemical analysis of his bodily fluids and solids.[57] Succi fasted there for forty days, in the presence of crowds of visitors paying a fee to see the peculiar oddity. But Succi's case was not an exception, and hunger artists became a frequent attraction in that particular context. Alexander Jacques, the French faster who was willing to emulate Succi in the same London context, was one of the witnesses to Succi's fasts in the Royal Aquarium.[58] In fact, Jacques also fasted the same year at the Aquarium for forty-two days, starting on 21 June and receiving visits all day at the West End Court.[59] Three years earlier, in 1887, the same palace of oddities had welcomed faster Francesco Cetti, whose audience included in that case not only physicians, but also journalists and members of the nobility. Cetti was introduced to

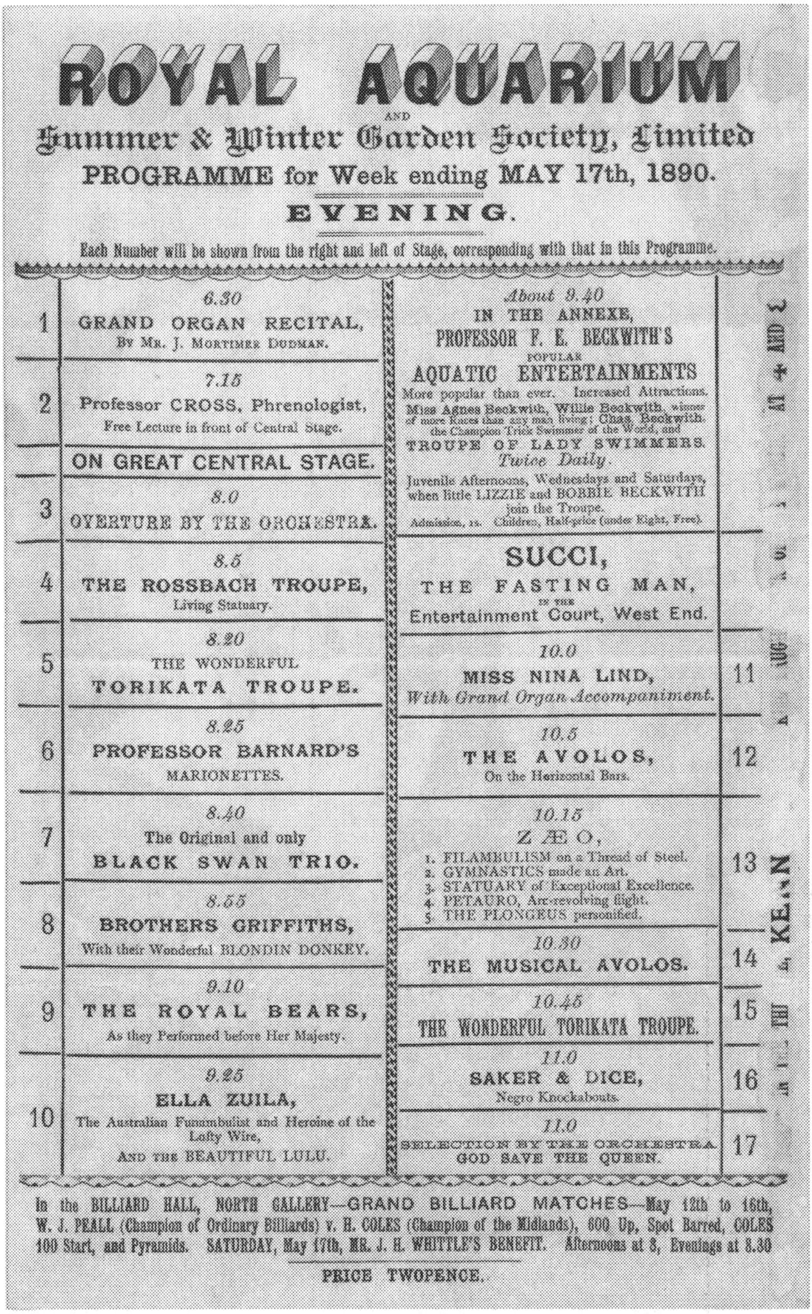
ROYAL AQUARIUM
AND
Summer & Winter Garden Society, Limited
PROGRAMME for Week ending MAY 17th, 1890.
EVENING.
Each Number will be shown from the right and left of Stage, corresponding with that in this Programme.

1 6.30 GRAND ORGAN RECITAL, By Mr. J. Mortimer Dudman.

2 7.15 Professor CROSS, Phrenologist, Free Lecture in front of Central Stage.

ON GREAT CENTRAL STAGE.

3 8.0 OVERTURE BY THE ORCHESTRA.

4 8.5 THE ROSSBACH TROUPE, Living Statuary.

5 8.20 THE WONDERFUL TORIKATA TROUPE.

6 8.25 PROFESSOR BARNARD'S MARIONETTES.

7 8.40 The Original and only BLACK SWAN TRIO.

8 8.55 BROTHERS GRIFFITHS, With their Wonderful BLONDIN DONKEY.

9 9.10 THE ROYAL BEARS, As they Performed before Her Majesty.

10 9.25 ELLA ZUILA, The Australian Funambulist and Heroine of the Lofty Wire, And the BEAUTIFUL LULU.

About 9.40
IN THE ANNEXE,
PROFESSOR F. E. BECKWITH'S
POPULAR
AQUATIC ENTERTAINMENTS
More popular than ever. Increased Attractions.
Miss Agnes Beckwith, Willie Beckwith, winner of more Races than any man living; Chas. Beckwith, the Champion Trick Swimmer of the World, and
TROUPE OF LADY SWIMMERS.
Twice Daily.
Juvenile Afternoons, Wednesdays and Saturdays, when little LIZZIE and BOBBIE BECKWITH join the Troupe.
Admission, 1s. Children, Half-price (under Eight, Free).

SUCCI,
THE FASTING MAN,
IN THE
Entertainment Court, West End.

10.0 MISS NINA LIND, With Grand Organ Accompaniment. 11

10.5 THE AVOLOS, On the Horizontal Bars. 12

10.15 ZÆO, 13
1. FILAMBULISM on a Thread of Steel.
2. GYMNASTICS made an Art.
3. STATUARY of Exceptional Excellence.
4. PETAURO, Arc-revolving flight.
5. THE PLONGEUS personified.

10.30 THE MUSICAL AVOLOS. 14

10.45 THE WONDERFUL TORIKATA TROUPE. 15

11.0 SAKER & DICE, Negro Knockabouts. 16

11.0 SELECTION BY THE ORCHESTRA. GOD SAVE THE QUEEN. 17

In the BILLIARD HALL, NORTH GALLERY—GRAND BILLIARD MATCHES—May 12th to 16th, W. J. PEALL (Champion of Ordinary Billiards) v. H. COLES (Champion of the Midlands), 600 Up, Spot Barred, COLES 100 Start, and Pyramids. SATURDAY, May 17th, MR. J. H. WHITTLE'S BENEFIT. Afternoons at 3, Evenings at 8.30

PRICE TWOPENCE.

Figure 2.4. Poster advertising Giovanni Succi's fast at the Royal Aquarium London in 1890. He shared the programme with all sorts of oddities.

the audience as a man who had been supervised by numerous medical authorities, in particular, at a Medical Congress in Berlin and under the oversight of the renowned prestigious physician and polymath Rudolf Virchow. Equally, the organisers announced that Cetti would be subjected to medical supervision in London during his fast.[60]

The exhibition of hunger artists in a freak show context raised controversy, but became a common practice in our period of study, a time in which even international exhibitions often welcomed displays of freaks, human zoos, and exotic tribes.[61] Interestingly, the amusement park on Coney Island had its first freak show in 1880, the same year Tanner fasted in New York. It was just a starting point for a long-term tradition of freak show exhibitions that lasted until the 1940s, and included oddities such as Lavinia Warren, the wife of General Tom Thumb, and Henry Johnson, who had originally worked in Barnum's circus.[62] Phineas Taylor Barnum,[63] the famous showman, businessman, founder of the Ringling Bros. and Barnum & Bailey Circus in 1871, and an endless exhibitor of all sorts of freaks, accompanied Succi to the numerous fasts the latter performed across the United States.[64]

Other palaces for oddities also welcomed hunger artists' performances.[65] The Sheffield Jungle held collections of animals, freak exhibitions, and circus spectacles. Primates were exhibited in the Darwin Villa close to hunger artist Victor Beauté's public fast and the Joy Wheel, a mechanical device with a circular platform visitors slid off. The Jungle also hosted other attractions such as Anita the Living Doll, Mary Ellen the elephant, polar bears, Tom Tallon the young lion tamer, Hans and Greta the chimpanzees, and Fritz the boxing kangaroo.[66] In 1907, Beauté had also fasted in Glasgow at Pickard's American Museum, Trongate, but the procurator-fiscal of Lanarkshire forbade the exhibition because of risks to the faster's health. Beauté was thirty-one years old and already had notable experience as a faster in Germany, the Netherlands, Belgium, and England.[67] In December 1910 and 1911, he also took part in the exhibition of freak shows and other oddities at the Sheffield Jungle. Although Beauté was enclosed in a glass cage and later examined by doctors from Sheffield University, his place in the show was similar to that of Osco and the snakes. Beauté fasted for twenty-nine days in the Jungle, with credentials of five years' experience as professional faster and with twenty-five performances under his belt. As reported in the Jungle Archive:

> Having fasted for 29 days, M. Victor Beauté on Saturday emerged from sequestration in his glass house at the Sheffield Jungle. He looked surprisingly strong and healthy, and told an 'independent' representative that during the five years he had made fasting his profession he had never had an easier trial. On the 25 previous occasions, he explained, he had not had such interesting surroundings.

Since his temporary prison had been sealed up on 1st April, his attention had been continuously attracted by his remarkable environment ... After bowing to a cheering crowd of people, and exchanging handgrips with numerous friends, Mr Beauté proceeded to the Sheffield University, where he added another 3 1/2 hours to his lengthy fast whilst undergoing a special scientific examination. In the evening he returned to the Jungle and took his first meal in the arena, in full view of an *intensely interested audience*.[68]

Again, Jacques, Cetti, and Beauté, as happened with Succi in the Royal Aquarium London, were perfectly integrated into the heart of that freak culture, which included freaks born that way (conjoined twins, missing limbs), freaks made that way (tattoos, long hair), novelty freaks (who gave unusual performances as hunger artists, sword swallowers, snake charmers), and fake freaks (frauds), but their potential for medical debates on food science spread far beyond the Royal Aquarium, like other oddities of their time.[69] In the freak show context, fasting brought contemporary actors to a fluid, cacophonous territory of fringe epistemology. Even in cases in which medical doctors tried to seal hunger artists in their laboratories to avoid suspicion and fraud, the freak show connivance was sometimes unbearable and inevitably influenced the way in which experts addressed the study of the causes of resistance to inanition. For hunger artists, the freak show arena was a useful extension of the public sphere. Even in view of the reluctance of a considerable proportion of the medical profession to accept the scientific value of these odd performances, the fact is that their growing popularity, never exempt from controversy, pushed doctors, journalists, science popularisers, and other varied witnesses to take the facts more seriously. Like the old lecture room in Barnum's American Museum, which originally hosted talks on natural history and natural philosophy, and later shifted to juggling exhibitions, magic, and freaks,[70] hunger artists crossed boundaries of orthodoxy and heterodoxy and played their cards to sell more tickets in a competition for entertainment and knowledge on hunger and inanition. Freak show venues were places that mattered epistemologically and reinforced the idea that medical interest in monstrosities could be extended to the study of inanition even beyond laboratory walls.

Commotion

Measurements and experiments by the medical committees in fasting places and private laboratories (see Chapter 3) seemed rather marginal in a jungle full of suspicion of fraud, unease about the painful experience of approaching the limits of human resistance, criticism against those who used public fasting for business and profit, and the never-ending tensions of different medical schools, all reflecting the virtues and miseries of hunger artists who appeared in contemporary media. In fact, public fasting performances provoked all sorts of reactions: from emotional feelings to rational criticism; from generalised scepticism to warm enthusiasm; from critical positions to provocative remarks; from scientific interest to media sensationalism.[71] It was in that messy public dimension that the plural views about the actual facts intersected and shaped the social truth about hunger. Details on the lives and performances of the best-known hunger artists can help us to assess how that commotion co-constructed contemporary knowledge on hunger and inanition in various ways.

The case of the pioneer faster, Dr Henry Tanner, is a good place to start.[72] Tanner was born in England in 1831. At the age of seventeen, he travelled to the United States and settled in Litchfield, Ohio, where he worked for years as a carriage maker. He then moved to Cleveland, where he began his medical studies in 1856, later joining the Eclectic Medical Institute of Cincinnati, where he graduated in 1859. He served as a hospital steward during the American Civil War, and later settled down as an 'eclectic' physician, in 1870, in Minneapolis. Before his startling appearance in New York, Tanner had performed his first fast in Minneapolis, in 1877, with some medical supervision, but certainly far from the crowded medical committees that took part in later fasts. Tanner conceived this first act of voluntary inanition as a way to cure a stomach inflammation, but with loose external medical supervision and no quantitative records of his vital signs. He counteracted criticism in the local press with statements about the generous aims of his painful experience for the sake of the therapeutics of the 'sick and suffering humanity', 'but not for notoriety, not for money'.[73]

In New York, Tanner's fast lasted from the 28th of June to the 6th of August 1880, in the Lecture Room of the US Medical College, at Clarendon Hall – 114 East 13th Street – under the supervision of a medical committee and that of *New York Herald* journalists. Far from philanthropism, the fast had a very specific business plan, which included bets, entry charges, the sale of pictures and postcards, donations, and financial support from private firms such as Justus von Liebig's meat extract company, which used the public debate on human inanition to advertise its own nutritional soups.[74] Tanner was also overseen by numerous visitors, watchers, and reporters, measured by instruments,[75] and supervised by members of the US Medical College, a group of medical students, and other physicians.[76] In fact, Tanner's popularity was considerable. On the eighteenth day, for example: 'a number of professional singers called and sang for the doctor, to his great satisfaction. He was also presented with many beautiful bouquets.'[77] Among his distinguished followers, it is worth mentioning Reverend Dr Deem of the Church of the Strangers, Dr Philip H. van der Weyde, Dr R. H. Gunn (a member of the medical supervision committee of Tanner's fast and his biographer), and Dr W. H. White, President of the County Homeopathic Medical Society (see Chapter 4).[78] But figures and profiles of his visitors were numerous and varied: theatre actors, women – a young lady from New Orleans who could not swallow solids or liquids of any description; other ladies who offered to marry Tanner; Ms Mollie Fancher (see Chapter 4) who sent him a letter declaring all her support – the famous American athlete Fred Englehardt offered Tanner $1,000 per week to deliver public lectures after his fast.[79]

As described in reports of his performance: 'Men and women began to file into the hall at 8 o'clock. Many of them were of peculiar aspect, representatives of various "advanced ideas", who came to participate in what each regarded in some way a triumph of the appreciation he or she represented. A majority, however, were mere curiosity-hunters.'[80] As the fortieth day approached, though, Tanner's pain and physical degradation made visits less advisable. On the thirty-fifth day, for instance, no letters or visitors – except members of the press – were allowed, and the box office for tickets was closed all day. Other letters considered Tanner to be a dangerous nihilist, a radical in scientific, social, and religious issues.[81] Some reports talked of 1,200 people cheering the end of his fast.[82] As the caricatures appearing on the walls show, Tanner became rather famous for his words of advice, his moderation in eating and drinking, and his common-sense recipes.[83] Nevertheless, a more disgusting commotion came from his enemies. Military physician Dr William Hammond (1828–1900), at the time president of the New York Neurological Society (NYNS), denounced Tanner's fast for supposedly being poorly supervised. The acting president of the NYNS, Dr Landon Carter Gray, addressed the editor of the *Tribune* in an excoriating letter against the fast:

> Dr Tanner is a man of sufficient intelligence to understand that he professes to be able to do what has never been done in the history of the world; and that, in order to prove that he alone among the many millions of countless ages can fast forty days, he must subject himself to *an examination as rigid and careful as human ingenuity can make it.* Even then there will be skeptics. *If he is not willing to submit to such test, it is illogical on his part to complain of the doubt expressed by scientific men.* Any person, well versed in sleight of hand, could feed himself or be fed with ease as the watch is now being conducted. Such prestidigitators as Hermann[84] and Houdini would grow fat in Clarendon Hall.[85]

Dr Gray's desperate attempt to enforce a clear boundary of authority by members of the established medical community was never satisfied.

It left room for speculation and the influence of other voices. The pattern was also repeated later in many European fasts. Stefano Merlatti's list of visitors to his hotel in Paris in 1886 was indeed impressive. Apart from prestigious medical doctors, he also received painters and photographers.[86] Likewise, during his fast in London in 1890, Succi had constant requests to sign autographs for the many people who flocked to see him and talk to him, often in long chats and interviews, which he combined with his readings, writing and a great amount of physical exercise.[87] In 1890, in New York, Succi welcomed numerous upper-class ladies during his fast, as described in the press: 'The ladies, by the way, have been his staunchest supporters and steadiest visitors. Many of them have come to see him time and time again, and one evening I counted seven carriages waiting for women who appeared to be accustomed to riding in carriages.'[88] Beyond women, 'actors and actresses ... scores of society' constantly 'dropped in to see what a starving man looked like, and of professional men there has been a regiment'.[89]

Many of Merlatti's enthusiastic visitors in Paris in 1886 ended up rather disappointed and even shocked by the torture and pain that these professional fasters suffered and by the supposed uselessness of their act in scientific terms. This was a general atmosphere that made it harder for medical doctors to pursue their experiments. The press reported that 'Many Parisians felt that the "affair" was becoming in some ways cruel, and the spectacle of this near-suicide almost barbaric. If there had been any hope of deriving a great scientific benefit from this fast, this advantage would have made one forget the humanitarian considerations.'[90] Dr Paul Loye, professor of physiology at the Faculté de Médecine de Paris, bitterly reacted against the 'theatre' of the Succi–Merlatti contest, with the visits, walks, rides, talks, and supervisions, which in practice he felt could easily push the whole endeavour over into fraud and distort the potential experiments, and again raising the issue of the doctors' collaboration in such a dubious performance.[91] In the 1880s Loye lectured on the brain and the nervous system, but was probably better known for his experiments on animal beheading. In Loye's view, the only reliable experiment in the case of hunger artists would have been a process of inanition until actual death.[92] In a similar way, Succi's fasts in London and New York in 1890 provoked bitter reactions in the press, which considered the performance to be particularly disgusting. In the case of New York: 'upon all general principles, the show is one which ought to disgust rather than interest. A ghastly, chalky face, feverish looking, glistening eyes, deeply sunken, great holes in the temples and cheeks, skinny, claw like hands and wrists – these are not beautiful things worth a quarter to go and see ... What is worse for the signor: there is an unpleasant prospect of his becoming ridiculous as well as hideous.'[93]

In 1888, in Florence, Professor Luigi Luciani's experiments on Succi's fast took place at the Laboratorio di Fisiologia[94] and were supervised by medical committees; 'with a considerable attendance of prominent Florentine *physicians*, with a large number of *medical students, lay citizens and newspaper reporters*, and two volunteers observing Succi day and night'.[95] In fact, Luciani was in favour of 'rigorous' experimentation on hunger artists, but opened it to the public sphere. Luciani argued that this was a good way to prove the absence of fraud, but at the same time it acted subtly as a way to gain social reputation. In Luciani's words:

> Although all these modern fasters have been accused of being jugglers and deceivers, throughout their fasts they showed constant decrease in weight, and *inspection by visitors was welcomed at all times*. They invariably invited medical attention, and some were under the closest surveillance; although ... the fasts were in every respect bona fide, yet we must acknowledge that these men displayed great endurance in their apparent indifference for food, the deprivation of which in a normal individual for one day only causes intense suffering.[96]

Witnesses of the long fasts and their social credibility as supervisors played a critical role during the performances. At least in France, the late nineteenth-century generation of science popularisers became relevant actors in the commotion in the public sphere. They often attempted to assess the credibility of professional fasters as part of their crusade against pseudo-science, superstition, and fraud. But, obviously, not all science popularisers shared the same view on these issues. Louis Figuier (1819–1894), for example, regarded his role as a science writer as a way to spread and legitimise the authority of experts, whereas socialist Victor Meunier battled for a popular science that was quite removed from '*science officielle*', with figures like the famous populariser of astronomy, Camille Flammarion – trying to find a balance between academic and popular science through his popular astronomy – very much in the middle ground.[97] Figuier perceived hunger artists, in particular the 1886 Succi–Merlatti contest, as useless for scientific inquiry, and as being closer to the realm of pseudo-science, charlatanry, fraud, or perhaps insanity. He summarised the case with a single devastating sentence: 'science has nothing to do with their adventure', which seemed to leave little room for a serious reconsideration of the facts.[98] Figuier was formally against spiritism, animal magnetism, and materialism, and placed Succi's and Merlatti's fasts close to that slippery terrain.[99] Paradoxically, however, after the tragic death of his son, he explored these uncharted territories with the publication in 1871 of *Le lendemain de la mort* (see Chapter 4).[100] At the other extreme, Victor

Meunier's fascination for popular science helped him to see hunger artists as potential bridges between the expert and lay scientific cultures, in a more flexible approach to the problem of the boundaries of pseudo-science. Meunier somehow recovered the medical eclectic tradition à la Tanner, which had allowed many physicians to use the professional fasters' performances as potential examples to reinforce different heterodox views, often closer to popular beliefs, on issues of health and diet.

The fiercest hostility against hunger artists came from science populariser, journalist, and aeronaut, Wilfrid de Fonvielle,[101] who bitterly discredited doctors and scientists who attempted to use hunger artists as objects for scientific research. Fonvielle also attacked the apothecaries for their collaboration in the control and examination of hunger artists' fluids.[102] He considered fasting performances to be 'secular' miracles or direct fraud. His book, *Mort de faim*, actually became a weapon against all our actors' 'theatre'.[103] He supported a radical rejection of all hunger artists' practices, since they could never be exempt from the suspicion of fraud.[104] Populariser Henri de Parville (1838–1909), who had become widely known for his novel *Un habitant du planète Mars* (1865), also endorsed the idea that hunger artists were useless for experimental physiology. He considered that their exceptionality had to be explained through cultural, educational factors or by a very special case of willpower, but their capacity to change scientific laws did not exist.[105]

Other critical voices came from average visitors, urban *flâneurs*, who contemplated the spectacle with bitter irony and discontent.[106] In a satirical report of the 1888 Barcelona Exhibition by comparison with that of Paris in 1889, journalist Carlos Frontaura gave voice to a family, Pedro, Manolita, and their two sons, in relation to their visit to Succi during his fast in the science pavilion:

> Don Pedro joined his wife and children, who were waiting for him, and went to visit Succi the faster, who occupied a room in the Palacio de Ciencias. The emulator of the American Doctor Tanner was there, guarded and watched over by a medical committee. 'Hey, [Pedro]', Doña Manolita said to her husband, 'who is that man? – It's Succi. – Chuchi? … And what's he doing? – He's not eating. – Is he sick? – No, you can see, he hasn't eaten for ten days, and he's walking, talking, jumping, handling the foil, laughing and nothing is happening. – Jesus Christ, what a clown! – He only takes the odd purgative, and every day his stomach is washed … – But he might eat something when no one is watching. – No, he doesn't eat. Can't you see that the supervision committee does not let him out of their sight for even one moment? … – What is this man trying to achieve by not eating? – Well, feed his family.'[107]

Again, the commodification of the fast and, as a consequence, the commodification of science itself seemed to be the conclusion of the lay visitors, whose thoughts and comments Frontaura reproduced with a clear satirical intention, probably accurately reflecting the unease that the presence of a potential charlatan in a science pavilion caused in the city.

The commotion also included disagreements between hunger artists, medical committees, and impresarios. Their mutual benefit, in terms of authority and the social prestige of their respective professions, did not hinder discrepancies in the procedures to monitor any potential fraud or hard negotiations for the sake of mutual trust. In 1886 in Paris, for example, Succi had important disagreements with his committee. Dr Barberi-Borghini, who had supervised a previous fast of Succi's in Milan and now introduced him into Parisian scientific circles, faced serious tension with other members of the medical committee. Dr Fernand Lagrange (1845–1909),[108] an expert on the physiology of sport – another field exploring the limits of the human body – considered that the whole experiment-performance needed a person (not an animal, like in Milan) to act as a blank sample. In addition, Lagrange demanded that the composition of Succi's liquor – a tremendous source of disagreement (see Chapter 5) – had to be submitted to the supervision of the Academy of Medicine. Barberi-Borghini's dislike of this change led Succi to resist the new rules bitterly. Succi decided to dismiss the official medical committee – paying the high price of losing the possible legitimation of his performance by the prestigious Parisian Academy of Medicine – and made an open call to any other doctor willing to act as supervisor and agreeing to follow the main guidelines of his doctor-manager.[109] Again, like in Tanner's case, reaching a consensus on what did actually happen during the fast seemed an impossible endeavour before the cacophony of different voices.

In Paris, in 1886, Succi signed a contract with the impresario Lamperti to establish the conditions of the fast. Lamperti had to pay 15,000 francs if Succi did not accomplish the following conditions: fast for thirty consecutive days, only drinking filtered water, Vichy and Hunyadi Janos, and only once some drops of Succi's famous liquor (on the first day) (see Chapter 5). During the fast, Succi had to walk, ride a horse, and do all sorts of exercise, swimming, fencing, gymnastics – all under the supervision of a medical committee – and he had to offer a public banquet at the end of his fast. Again, in a struggle for authority and legitimation, the Paris Academy of Medicine asked for stricter supervision, which he did not accept. Since Succi drank the liquor on the eighth day, Lamperti did not pay the agreed sum, and the conflict went to court.[110] Succi's absence from the Eden theatre – one of the sites chosen for the performance – also contributed to Lamperti's disappointment. As described in the press: 'Succi the faster is really unlucky. After having deprived himself of food for the thirty days agreed upon, having put his stomach to the most perilous of tests … he hoped and had the right to hope to receive the prize for his voluntary abstinence, that is to say fifteen thousand beautiful francs which an impresario, Mr Lamperti, had undertaken to pay him, subject to certain conditions.'[111] In practice, the conditions of the Succi–Lamperti contract shaped the way in which the fast was organised and appropriated later by the different witnesses.

In the case of Merlatti, there is no evidence of an impresario, but the conditions of the fast established by the medical committee, led by Drs Monin and Maréchal, were again very strict and detailed:

(1) Merlatti had to take no food and only filtered water during the whole fast;
(2) Two members of the medical committee would supervise Merlatti day and night;
(3) The committee would name one overseer for the day functions and the other for the nights;
(4) The first watcher would be replaced, at dinner time, by a member of the committee;
(5) The second watcher would keep Merlatti's chamber locked the whole night;
(6) Merlatti undertook to accept all restrictions imposed by the committee;
(7) Every morning a report on Merlatti's condition would appear with details of weight, pulse, temperature, urine composition, and water absorbed;
(8) Pharmacist Edmond Vasseur would be in charge of the urine and faeces analysis;
(9) The committee would accept any demand from Merlatti to stop the fast, and provide some advice on how to proceed in terms of his feeding;
(10) If severe symptoms, ones risking Merlatti's health, appeared, the committee would have the right to forcibly stop the fast, and the faster would accept the conditions.[112]

Again, in this case, public complaints and disputes were raised in the last days of the fast. Doctors pushed him to stop the experiment for the sake of his health, and several newspapers denounced the cruelty of pushing a human body to its limits.[113]

In 1888, in Barcelona, three days before the beginning of his public fast, Succi addressed the Sección de Ciencias Exactas y Naturales of the Ateneu Barcelonès, an influential cultural institution in the city. Again, seeking scientific recognition, he tried to convince the members of the Sección that his experiment was based on solid physiological principles, and showed his numerous recommendation letters from foreign doctors who endorsed his credibility in former fasts. Two distinguished members of the Sección, Javier de Benavent and Salvador Badía, also defended eclectic medical practices. Benavent practised homeopathy (see Chapter 4), and Badía was a member of the Catalan Vegetarian League and close to theosophy.[114] However, other members of the medical committee were well-established medical doctors, some with positions at the Faculty of Medicine of the University of Barcelona.[115]

Relevant personages, who supported public fasts for their own professional interests, also established regular collaboration with newspapers and periodicals. In Tanner's fast, for example, the press played a crucial role. In particular, Tanner's controversy with Dr Hammond about the reliability of the conditions of the fast, or their disagreements about the scientific nature of fasting girls (see Chapter 4), could not be properly understood without the extended battlefield of the controversy in the media.[116] In a similar vein, Dr Philippe Maréchal, editor of the periodical *Voltaire*, and Dr Ernest Monin, editor of *Gil Blas*, were both deeply involved in Stefano Merlatti's fast in Paris in 1886.[117] Science populariser Victor Meunier belonged to the supervision committee of Merlatti's fast, but Meunier was also the scientific editor of newspaper *Rappel*.[118] In other cases, such as Succi's fast in New York in 1890, reporters from the *New York Herald* took a very prominent role in the organisation of the fast, in its dissemination in their newspaper, and in the supervision committees.

In fact, the press often delegitimised the authority of the medical committees of public fasts and tried to have their say in the whole debate, adding more commotion to public controversies. This was, for example, the reaction of the *Manchester Courier* after Succi's fast in New York in 1890, when the pompous closure of the public ritual contrasted with a gloomier vision of the event in the English newspaper in the following terms: 'The whole thing has been a failure in New York. The public have not interested themselves in [Succi's] task, and a small notice has been taken of it in the newspapers. Some journals have denounced the exhibition as disgusting, and its alleged scientific interest as all rubbish. The fast, however, was a genuine one, Succi having being constantly watched by reporters and doctors.'[119] In spite of the public controversy in the daily press, or perhaps because of it, the prestigious Bellevue Hospital Medical College sent eight of its physicians to take constant note of the progress of Succi's fast in New York,[120] to measure weight, temperature, pulse, respiration, and physical strength, to test his lungs, and to report on the water drunk and the general condition of the hunger artist. In particular, Dr Bettini di Moise became Succi's own physician during the fast and studied him even before the beginning of the inanition to compare his medical condition afterwards.[121]

The *New York Herald* reporters also became relevant actors within the whole performance. They stressed the impressive strength Succi showed during the fast which, beyond any suspicion of fraud, was legitimised by the medical committees: 'When Succi went through his last fast, at the Royal Aquarium in London, the physicians issued daily bulletins of his condition, just as the Bellevue physicians are doing now. If anyone doubts the extraordinary maintenance of strength which Succi asserts he has only to look over these bulletins.'[122] The *New York Herald*'s coverage of Succi's fast in New York was an impressive, thoroughly documented endeavour. From September to December 1890, more than thirty articles related to the event appeared in the newspaper, with appealing titles such as: 'Is Succi's Secret Will Power or Paranoia?', 'Empty Bread Baskets as a Steady Diet', 'Succi Wins His Race with Starvation', 'Doctors Wonder at the Succi Mystery', and 'Ups and Downs of Succi's Hunger Spell', among many others.[123] In an interview the same year, Succi addressed himself to the readers in the following terms: '"Look at that pile of doctors' reports", said Signor Succi. "I had them all bound up with the certificates and diplomas awarded to me on various occasions … Here are some photographs taken of me during different stages of my Milan fasting experiment; and this is a group of journalists and doctors who formed my surveillance committee in Brussels."'[124]

Nevertheless, accusations against professional fasters were frequent and common. They were often perceived as fraudsters, conjurers, and charlatans, who used the daily press to gain false prominence in the public sphere. Many firmly believed that in their performances hunger artists cheated by persuading their overseers to illicitly provide them with food during their public performances, and merely pretended to resist inanition for more than forty days.[125] In fact, Succi's own statements, substantiated by newspaper reports, showed that he carried out a large number of fasts which were not scientifically supervised. From Tanner's times, laboratory walls seemed therefore more permeable than expected to the public commotion produced around the issue of hunger artists. In 1895, fifteen years after Tanner's fast in New York, Dr John J. Rees, Professor of Medical Jurisprudence and Toxicology at the University of Pennsylvania, described the legal problem in the following terms:

Many cases of voluntary fasting, which have claimed the notice of the public during the past years, have proved, on close examination, to be *deceptions*, food and drink having been supplied surreptitiously to the individuals concerned ... The notorious Dr Tanner, of our own country, undertook, for a consideration, to perform the feat of a forty days' absolute fast, in New York, in August 1880, and, to all appearance, he accomplished it ... This case was not under very strict medical supervision, and the doubt as to its genuineness seems to be confirmed by *the fact of his voracious feeding on the completion of the fast, unattended by any bad effects, which is contrary to the general experience of others who have been deprived of food for a length of time*. Since the above case, several instances are recorded of voluntary fasting for periods varying from several weeks to some months; but *there is nearly always some uncertainty as to absolute fasting*.[126]

Popularisers, impresarios, medical committees, journalists, and the public at large spread disagreement, controversy, and speculation, which apparently weakened scientific authority, but left more room for doctors and hunger artists to actually gain social visibility. It became a blurred but genuine *way of knowing*,[127] with no clear boundaries between the descriptive nature of the fasts, the analysis of its various kinds of elements of inquiry, and the 'right' or 'wrong' way to run experiments on them (see Chapter 3), with all this resonating in the commotion of heterodoxy.

***

As Kafka's story reflects, hunger artists suffered the anxiety of the inevitable tensions of their time, between creating a supposedly autonomous art and the almost unavoidable influence of a commercialised context.[128] From its religious origins and the mystical tradition of scepticism, and even from its use as a therapeutic practice, fasting had become a new commodity, a spectacle to be sold and bought in the marketplace. It represented a deep shift in the cultural (and political) function of self-starvation (see Chapter 6). From that perspective, hunger artists appeared amidst the tension between high art and mass culture, between individual creativity and the inevitable influence of the masses (urban visitors but also impresarios, doctors, journalists, and politicians), in between scientific expertise and a heterogeneous public sphere, in the appropriation of their performances.[129]

In 1907, German writer Robert Walser published the novel *Geschwister Tanner*, which used the name of the pioneer American faster. Walser's life – his loneliness and voluntary marginality, his profile as a 'starving poet' more concerned about creative writing than eating – seemed to have influenced Kafka. It resembled the profile of many professional fasters who lived on the fringes of bourgeois conventionalism, endlessly travelled as wandering charlatans, and praised personal freedom, austerity, sacrifice, and self-discipline.[130] In 1912, the renowned nutritionist Francis Gano Benedict published a paper titled 'An Experiment on a Fasting Man', in which he noticed the inevitable influence in newspapers and magazines of the whole endeavour. He regretted that: 'The mass of data will require several months for complete and verified computation. Newspapers and magazines, actuated only by the sensational element, have used every means to secure advance statements, and in some instances have issued "faked" statements, regarding this experiment.'[131] Benedict regretted that, even when hunger artists were supposedly 'pure' objects of scientific experimentation or agents of artistic creativity, they did not escape from the logic of the marketplace.[132] His annoyance about the commotion in the public sphere, which in his view distorted the efficiency of his 'scientific ivory tower' at the Carnegie Institute of Nutrition, fits well with the attempts of many professional scientists to 'isolate' their work from 'interference' and 'public commotion', which was most probably an unachievable endeavour.

But Benedict was only one of the actors in a crowded play. Fasting performances with their spatiality, overseers and witnesses constructed a complex social truth out of those controversial events. In between science, spectacle, fraud, scientific interests, and business strategies, in between the *episteme* and the *doxa*, and in spite of the endless disputes about the credibility, objectivity, and rationality of facts, they constituted very rich cultural packages that circulated at different levels. Medical committees, managers, impresarios, journalists, science popularisers, and hunger artists themselves were keen to build up narratives of credibility, which could easily shift towards the edges of 'irrationality' in freak shows, in places such as dime museums, the Royal Aquarium London, and the Sheffield Jungle. These cacophonic opinions brought considerable public commotion to the fore and made a potential construction of a common agreement on public fasting harder, or even impossible, to achieve. Nevertheless, the commotion in *the land of the hunger artists* gave a voice to hitherto unknown historical actors – from hunger artists themselves, to impresarios, journalists, and interpreters – who inevitably influenced debates on issues such as the limits of human resistance to inanition, the therapeutic uses of fasting, and the non-material explanations of the functioning of the human body. The focus on a more popular, eclectic approach to medicine and health offered a new way to revisit reluctance about animal experimentation; it drove an unexpected interest in fraud in epistemological terms as well as the need to promote new disciplines and medical specialities to face the complexity of terms such as hunger, appetite, fasting, inanition, and starvation.

In the next chapter, I shall describe the 'scientific' endeavour that many physiologists had in mind when they attempted to escape from the public commotion and approached hunger artists as idealised objects of experimentation.

3 Experiments

Doctors who studied hunger artists in their laboratories were often skilful at animal experimentation. Luigi Luciani, for example, was widely known for his study on Succi in Florence in 1888, but three years later he published his famous study *Il cervelletto* as a result of the removal of a dog's cerebellum and the systematic study of the animal's behaviour and physiological evolution for days.[1] Although the study of the encephalon went beyond the strict domain of hunger and inanition, the role that Luciani attributed to the nervous system in prolonged fasts linked the animal organ and Succi's case. In the 1880s, physiology textbooks discussed where the feeling of hunger was physically located in the body. Doctors such as Austin Flint (1812–1886) at the Bellevue Hospital in New York – a prestigious medical institution that had supervised Tanner's fast in 1880 – suggested that nerves connected to the brain, or more specifically to the encephalon, controlled the feeling of hunger. In 1886, François-Achille Longet (1811–1871), a renowned anatomist and professor of physiology at the Paris Faculty of Medicine, obtained similar results when he discussed the limitations of a strict materialist philosophy to explain Merlatti's resistance in his contest with Succi.[2] The same year, Dr Jean Baptiste Laborde (1830–1903), the president of the French Société d'autopsie, assessed the role of water in dogs when subjected to a total absence of solid food. Using two dogs of the same breed, Laborde concluded that a generous ingestion of water considerably delayed the death of one of the dogs (forty days) when compared to the other animal which was completely deprived of liquid. He extrapolated his experiment with animals again to the case of Merlatti, whose resistance to inanition for fifty days could perhaps be explained, he thought, by his slow ingestion of water. With hunger artists, the boundary between animal and human experimentation seemed therefore to vanish.[3]

Measurements of vital signs and bodily fluids in animals up to death were common in physiology laboratories.[4] Decades before the appearance of professional fasters, animal experimentation on processes of fasting, inanition, and starvation had a long tradition. At the end of the

1880s, in his courses of physiology in the Paris Faculty of Medicine, Charles Richet described in full detail the resistance of different animals to inanition and the associated medical research. Dogs, cats, rabbits, horses, pigs, rats, eagles, ducks, pigeons, and chickens were associated with prestigious names such as Luciani, Bufalini, Chossat, Bernard, Magendie, Colin, and Richet, the latter placing himself in the pantheon of these experimenters.[5] Alongside these experiments, in scientific journals and in the general press many doctors questioned the supposed additional value of experiments on humans when compared with those on animals.[6] The case of hunger artists added more nuance to the animal–human experimental question.

In the mid-nineteenth century, medical authorities such as Charles Chossat (1796–1875) established the ratio of weight loss that put the lives of different animals at risk when deprived of any kind of food except water.[7] But animals did not behave as some hunger artists did. They did not ride horses, practise fencing and take exercise, or talk with visitors. Animals just remained enclosed in cages, suffering day after day, until their death by inanition, in experiments that apparently seemed easier to standardise. In addition, humans, hunger artists in our case, as objects of scientific experimentation were often suspected of fraud and required extra surveillance and control, which could potentially damage the prestige and authority of experimental physiology. In 1886, the French medical journal *Le progrès médical* pointed out that: 'It is precisely because the experience seemed odd and paradoxical that it demanded more careful and stricter surveillance.'[8]

In spite of all this reluctance, others thought that hunger artists could offer a golden opportunity to reinforce experiments on humans and to increase contemporary knowledge on the resistance of the human body in a process of prolonged fasting. Philippe Maréchal, for example, one of the two doctors (with Ernest Monin) who promoted Stefano Merlatti's fast in Paris, had been educated at the Paris Faculty of Medicine, but became president of the Société protéctrice des animaux. Maréchal positioned himself in public against bull-fighting and vivisection and opposed animal experimentation of any sort. He also joined the Société française contre la vivisection, which was founded in Paris in 1882, following the lead of Dr George Hoggan, from London, and with Victor Hugo as its honorary president.[9] In Monin's view, experiments with hunger artists could avoid animal torture, as well as providing a closer look at the subtle frontier between inanition, starvation, and final death, which was harder to determine in animals.[10]

In 1890, Dr Henri Labbé, the laboratory director of the Paris Faculty of Medicine and an expert on nutrition and diet, described the challenges of hunger artists as subjects of experimentation in relation to animals. He pointed out that: 'The physiological phases and episodes of human fasting are less known than those of animals ... Fortunately, *professional fasters* have been able to persist in abstinence for much longer for the sake of gain and publicity. At the same time, they have agreed to undergo the necessary scientific research about themselves and their eliminations.'[11] Following Richet, Labbé noticed that during inanition nervous tissue in humans was more resistant when compared with animals, so in his view human experimentation was particularly needed to shed more light on the physiology of inanition.

In the 1880s, therefore, the old experimental tradition with animals faced a new challenge in relation to human experimentation, to self-experimentation, and, more particularly, to the scientific value of experimenting with hunger artists in the process of prolonged absence of food. Physiologists attempted to 'isolate' the professional fasters and confined them in the supposed 'quietness' of their laboratories. But to what extent could living human beings, and these peculiar characters in particular, be reliable objects of scientific inquiry?[12] In a process of inanition, could they add some extra epistemological value to the experience already acquired with animals? Were those charlatan-like characters disciplined objects of inquiry, or perhaps just a risky endeavour for the professional reputation of the medical experts involved? These were not easy questions to answer for contemporaries in our period of study.

Because they were accustomed to being confined in freak shows and public fairs, being enclosed in laboratories and sophisticated respiration chambers did not make a priori a big difference in terms of the fasters' wellbeing and capacity to adapt, so, according to many doctors, it was worth investing in the study of fasters under the constraints of the laboratory culture of the experimental physiology of the time. It was, though, clear that hunger artists sought medical supervision and experimentation for the sake of legitimation of their controversial performances and an increase in their financial gains. Also, this supervision usually implied the 'objective', expert measurement of the composition of their urine, faeces, and blood as a useful weapon against frequent suspicion of fraud.[13]

The golden period of hunger artists went hand in hand with a physiology that had reinforced its experimental character. As described in the following sections of this chapter, the prestige of new instruments legitimised the value of quantification and the 'objectivity' of diagnostic methods, through precision measurements, numbers, graphical inscriptions, and a growing reductionism of the functioning of the body to the laws of physics and chemistry.[14] That pattern could also be applied to the study of the human experimentation of prolonged fasting. In 1888, after Succi's stay in Luciani's laboratory in Florence, the British medical journal *The Lancet*, for example, reacted optimistically, pointing out that: '*The result was watched with special interest by Italian surgeons and physicians* anxious to get some scientific evidence of the time during which the cyclo-poietic viscera can remain undisturbed by the ingestion of food – a point of the highest importance in abdominal surgery or in bowel haemorrhage from perforation.'[15] In 1907, Dr Georges Linossier (1857–1923), professor of chemistry in Lyon, reported on the usefulness of hunger artists for research on the treatment of stomach ulcers and referred to Succi, Cetti, and Breithaupt as ideal cases of study, who had helped doctors to overcome their former animal experiment culture. Linossier regarded professional fasters as patients subjected to a water diet, by analogy with the therapeutics of absolute stomach rest, and water ingestion subcutaneously or through the bowel.[16]

Nevertheless, the hope that hunger artists would prove useful for medical purposes was shattered by more pessimistic diagnoses. In 1886, after Succi's fast in Milan, two medical students, Carlo Garampazzi and Giuseppe Raineri, who had joined the medical supervision board, reported their physical–pathological considerations and expressed their rough scepticism about the scientific value of the faster in the following terms: 'ruining one's health for a whim or to please the public under the guise of the most poetic man in the world is something that may arouse some curiosity, perhaps may make sellers of edible products fear for their future, may encourage the learned to conduct research, to reconcile the idea of a new Prometheus who steals a new flame, may convince the public about the Americanism that is penetrating Italy: it is however a dangerous method of speculation, a painful story ... to kill oneself to live'.[17] In a similar vein, Tommaso Venanzi was censorious: 'Succi, with his thirty-day fast, threatens to astound not only ignorant people, those who are mediocre or superlatively educated, but even scientific men of the toga.'[18]

After Succi's visit to London in 1890 for his 45-day fast, the *British Medical Journal* considered that public performance was unnecessary from the point of view of the physiologist: 'The practical physician and his patients will derive no benefit from it, and the general public can derive nothing other than harm from witnessing for a prolonged period a fellow creature in suffering.'[19] Results obtained in animals had to be extrapolated or adapted to humans, but psychological factors of disgust and indigestion, or fatigue from the tiny spaces in which they had to often spend days, made the experiments more complicated and uncertain.[20] Moreover, self-experimentation added extra complexity to the whole endeavour. Following the pioneering example of Henry Tanner in the 1880s, some doctors opted for that practice – in our case, self-prolonged fasting – as a safer method to control all the variables and to avoid subjectivities of the agency of other humans. In 1916, Anton Julius Carlson (1875–1956), recently appointed as chairman of the Department of Physiology at the University of Chicago, described the specific scientific value of self-experimentation on hunger through the functioning of the stomach in the following terms: 'The work on man was carried out on the author and one assistant, in such a way that the gastric hunger contractions and the subjective hunger sensation were recorded, day and night, during the entire starvation period. In animals below man we can, of course, record the gastric hunger contractions only, as we have as yet no means of determining the intensity or quality of the sensations caused by these contractions.'[21]

Hunger artists and their periods of resistance to inanition constituted an experimental system that this chapter seeks to explore through different analytical levels.[22] First, the quantifying, measuring nature of experimental physiology benefited from human fasting experiences as another example of the usefulness of new instruments to describe the functioning of the body in the absence of food. Claims of 'objectivity', even the mathematisation of the body's variables during inanition, reinforced the creation, use, and spread of new, sophisticated scientific instruments, from the *kymograph* to the respiration calorimeter. Secondly, hunger artists added new challenges to the experimenter–experimentee relationship in physiology laboratories, in terms of discipline, reliability, and fraud control and shifted the boundaries between material and spiritual explanations of ambiguous concepts such as hunger, appetite, inanition, and starvation. Finally, the absence of food provided new opportunities to study human metabolism and contributed to new quantitative approaches to the emerging science of nutrition as a new medical speciality. Around 1900, data on hunger artists during their fasts filled physiology and nutrition textbooks and became relevant for medical experimentation. Nevertheless, as in other experimental systems, the boundaries between the *epistemic*, the *technical*, the *social*, and the *institutional* were fluidly defined.[23] Although the physiologists' dream of isolating their experiments from the 'commotion' of the public sphere (see Chapter 2) was more a wish than a reality, their multiple attempts to enclose hunger artists within the discipline of the laboratory surely deserve further attention.

Instruments

From the mid-nineteenth century onwards, in a context of trends towards growing quantification and mathematisation of bodily functions, new scientific instruments populated the laboratories of experimental physiology and brought a new aura of modernity and progress in an optimistic attempt to extrapolate methods of physics and chemistry to medicine.[24] The invention of Carl Ludwig's (1816–1895) kymograph (wave writer) in 1847 represented a milestone for the new experimental culture, just like Justus von Liebig's *Kaliapparat* became, also in the 1840s, a key object for the legitimation of animal and vegetable chemistry.[25] The kymograph consisted of a revolving drum wrapped in paper, with a stylus that recorded changes in motion and pressure. Ludwig used it to measure blood pressure, but with its multiple variations it soon became the icon of the Leipzig Physiological Institute (founded in 1865) and represented well the successful materialisation of the new experimental physiology. The instrument reinforced Ludwig's materialist agenda and his commitment to a physical–chemical reductionist explanation of the functions of the human body.[26]

Quantification trends and a new instrument culture spread from Germany across Europe and later to the Americas as a new way to measure breathing, speech, body temperature, muscular strength, blood pressure, digestion, and more.[27] Once the new instruments, the kymograph but also the sfigmograph, the dynamometer, the pneumograph, and the respiration calorimeter, among many others, reached physiology laboratories, mechanical, electrical, and photographic recording became a new medical language, with a strong emphasis on the visual representation of experimental results.[28] The new instruments contributed to the growth and prestige of Ludwig's school, which included luminaries such as Angelo Mosso (1846–1910) and Carl von Voit (1831–1908), both involved in research on inanition. In the 1870s, Mosso stayed in Ludwig's laboratory and developed a new apparatus, the ergograph, to quantify human fatigue through the recording of the force and frequency of finger flexions (see Chapter 6).[29]

The new instruments acted as powerful mediators for the legitimation of expert doctors, but also provided extra credibility to fasters. The apparent plus of objectivity and quantification of pulse, blood pressure, muscular strength, and temperature, along with the chemical analysis of urine and the determination of the amount of nitrogen consumed, helped physiologists to transform subjective feelings of hunger and appetite or painful processes of inanition and starvation into 'reliable' ciphers, graphics, and numbers. In Paris in 1886, Merlatti's medical board referred, for example, to several instruments as key indicators of the 'scientific' supervision of the fast. As reported by Merlatti himself, 'these gentlemen began their studies: one dried my mouth by removing saliva with a small pump; another burned my eyes by shining a stream of light into each eye. But the most curious were the dynamometry experiments, consisting of making me pull on a rope, to the end of which was attached a steel apparatus forming a Roman balance.'[30] To supervise the fasts, Merlatti's board also used a thermometer, a spirometer, a dynamometer, a spirograph, graduated beakers, an ophthalmoscope,[31] and the spirograph strokes of Dr Oscar Jennings, the correspondent for *The Lancet* in London.[32] Merlatti's tactile sensitivity was measured with another instrument: Weber's aesthesiometer.[33]

Luciani's study of Succi in Florence in 1888 is another valuable example. In his book *Fisiologia del digiuno*, issued after carefully studying Succi for thirty days of fasting in his laboratory, Luciani quantified the consumption of human tissue, the daily deficits, chemical and gas exchanges, and temperature changes in the absence of food ingestion, and compared these figures with those obtained from animals.[34] He made extensive use of devices such as the pedometer, which measured the number of steps taken by Succi during his walks; the sphygmomanometer that measured his pulsations and blood pressure, to be later represented graphically using a sphygmogram; the dynamometer, which quantified Succi's muscular force, graphically represented in a dynamograph, an adaptation of Mosso's ergograph to quantify fatigue (Figure 3.1(a)).[35] During Succi's fast, fatty and muscular tissue disappeared much faster than bone and nerve tissue. Changes in his blood (red cells, white cells, and haemoglobin) were studied and quantified with the haemometer.[36] At the next stage, Luciani mathematised the daily deficit during inanition and described it as a hyperbolic curve.[37]

Luciani presented a graph, which synthesised different functions of Succi's body during inanition: temperature, breathing frequency, arterial pulsations, and arterial pressure (Figure 3.1(b)).[38] He also attempted to quantify the material exchange (inputs and outputs) involved during the fasts. Figures included products of urine (nitrogen, oxalic acid, chlorine) and carbon dioxide exchange through the lungs. To quantify gas exchanges during breathing, he built a pneumograph (or spirograph) to record the speed and force of the chest during breathing and a gasometer to calculate the carbon dioxide/oxygen ratio.[39] Luciani placed Succi in a respiratory apparatus (Figure 3.2) that consisted of a kind of elastic rubber funnel suitable for the shape of the dental arches, introduced between them and the lips, which act as obturators; a large glass tube which forks into two, joined by means of flexible rubber tubes, with two valve devices, which collect the saliva that could potentially drain from the mouth. The inspiratory valve contained a 50 per cent caustic potash solution and was joined by two tall cylinders filled with pieces of pumice soaked in a solution of potash and a few pieces of solid potash. The expiratory valve contained distilled water and was joined to a rubber cloth

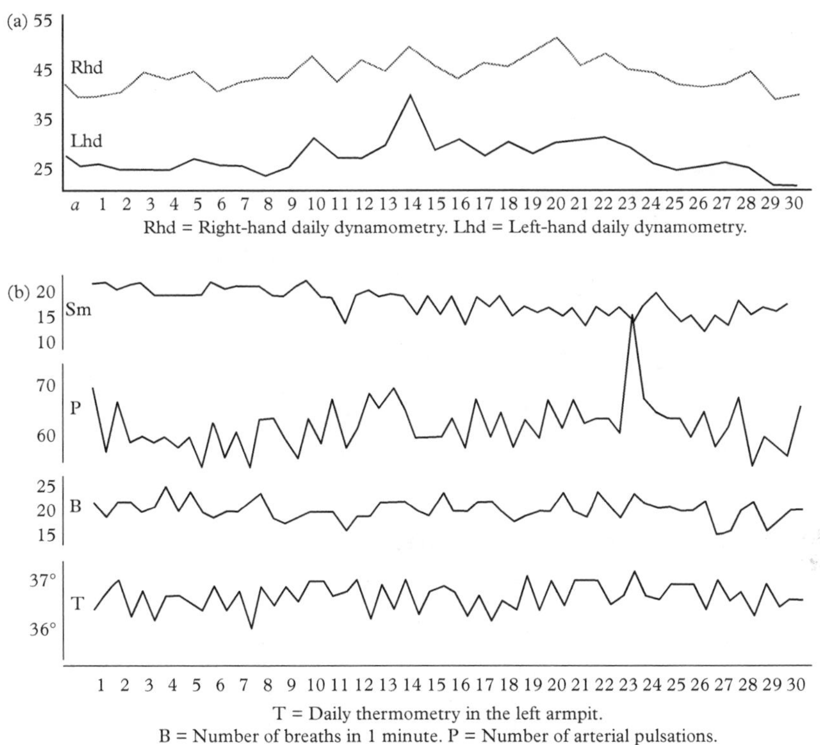

Figure 3.1. (a) The measurement of Giovanni Succi's right- and left-hand muscular strength during the fast, using the dynamometer. (b) The measurement of body temperature, breathing frequency, arterial pulsations, and artery pressure during inanition.

bag with a capacity of 100 litres, designed to collect all the air, saturated with moisture, exhaled in 10 minutes of breathing in the device. A pneumograph applied to Succi's chest traced in a rotating cylinder the number and excursion of the respiratory acts that took place in the ten-minute duration of each research period, so that the gasometrical analysis was always made of a homogeneous mixture of the air exhaled by Succi in half an hour of breathing. Finally, the pneumographic trace taken at the same time was a useful document of the regularity, depth, and average frequency of the respiratory rhythm throughout the experiment.[40]

Luciani's respiration apparatus should be placed in the context of the appearance of several sophisticated instruments labelled as 'respiration calorimeters', which were used in physiology laboratories. From the

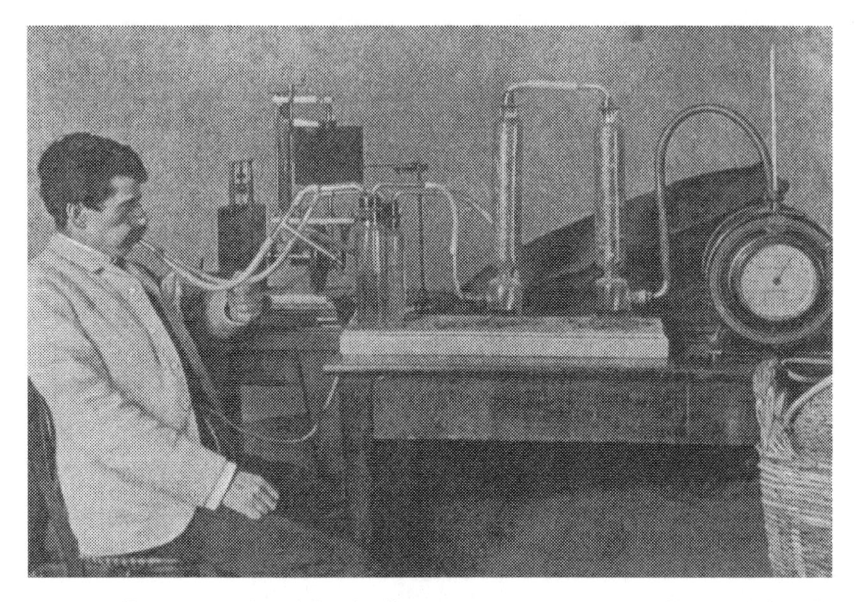

Figure 3.2. Luigi Luciani's apparatus to measure Giovanni Succi's respiration exchange during inanition.

'balance of matter' of the times of Lavoisier at the end of the eighteenth century, an accurate, quantified study of the intake–output matter in the human body and, more generally, a balance of the natural transformations of the animal organism became an enormous scientific challenge, which transcended academic disciplines, experimental methods, and philosophical doctrines.[41] As described in more detail in the next sections, the carbon, hydrogen, oxygen, and nitrogen that the animal, or the human being, acquired in the food eaten had to equal its excrement (in urine and faeces) and the gases involved in the respiration exchange. The quest for a reliable instrument to fulfil that difficult task occupied the efforts of many doctors and scientists, with the case of inanition in animals and humans being an ideal opportunity to refine data and theoretical considerations.

Luciani's experiments on Succi's tissues and functions during his fast allowed him to draw some general conclusions on the physiology of inanition, which were first published in 1889 in Italian and translated into German a year later.[42] With his set of instruments, Luciani had proved that, unlike other contemporary hunger artists, Succi's physiological functions remained rather stable during his fast, with a very low daily loss of proteins and a considerable absence of pain. Breathing, blood circulation, muscular and nervous activity, coenaesthesia (general wellbeing), and body temperature all remained quite stable, with only a gradual decrease in the concentration of nitrogen, sulphur, and phosphorus in the urine. In Luciani's view, only individuals with considerable amounts of stored oxidisable matter and slow consumption speeds could resist such long periods of inanition. Succi was an ideal and exceptional example of such a person.[43] Succi's mature age, his inherited slow metabolism, and his robust muscular and fatty tissue were the three crucial factors that explained his physiological stability and impressive resistance during the fasting. To those explanations, Luciani added Succi's skill at regulating the amount of water he drank and the role of mineral salts dissolved, which, because of their capacity of chemical combination with organic tissues, could be truly considered as nutrients during the fast (see Chapter 5).[44]

Luciani was convinced that the quantified study of Succi's fast in his laboratory would provide useful new data for further research on energy consumption at rest, as well as on muscular, mental, and nervous labour and psychological state in humans. He developed a general theory of inanition in three stages: *hunger* (a brief, painful period), *physiological inanition* (a longer period – twenty to twenty-five days in which consumption and temperature decrease, but the body functions stay quite normal), and *morbid inanition* or *crisis* (which precedes death, with hypothermia, vomiting, diarrhoea, and collapse). Through the inanition process several types of tissue also struggle to survive in the body and losses varied: 93 per cent adipose, 62–71 per cent (pancreas, liver, blood), 42 per cent (muscles), 27–32 per cent (kidneys, lungs), 17 per cent (bones), and 2 per cent (nervous system). Therefore, the inanition process implied a huge loss of fatty tissue, a very quick disappearance of carbohydrates during the first days, and a loss of proteins from the first two to three days onwards. In addition, Luciani's future textbooks relied on the experiment with Succi in Florence, which contributed to his later standard description of the physiology of fasting.[45] Following Tanner's path, Luciani believed that experiments with hunger artists could provide new data on the therapeutics of diet and its application to several illnesses.[46]

After Luciani's early achievements and the spread of new physiology instruments, the real challenge was to optimise the respiration calorimeter. Borrowing experimental skills from physics and chemistry, many doctors ran a series of intake–output studies.[47] The challenge was therefore ensuring accuracy in the measurement and quantification of the balance. It took a long time to calibrate and stabilise the respiration calorimeter, as well as to optimise the gas analysis methods and control of all variables. Attempts to accurately measure the intake and output of matter and energy in the animal organism were part of the scientific programme of many physiology schools. In a long-running contest for a reliable instrument and a reproducible experiment, the different versions of the calorimeter, in which hunger artists were placed under the strict control of the experimenter, circulated widely and were labelled with prestigious names in experimental physiology.[48]

In 1862, Carl von Voit, at that time professor of physiology at the University of Munich, sought the collaboration of Max von Pettenkofer, head of the Institute of Hygiene and Medical Microbiology of the city,[49] to design a respiration apparatus which could accommodate a dog. But Pettenkofer foresaw the advantages of designing the apparatus large enough to accommodate a human and built a chamber that was ventilated by pumps to draw air from outside.[50] After contributions by von Voit and Pettenkofer, others followed their lead. Von Voit's pupil, Max Rubner, a key name in the quantification of the metabolism of fasting, the creator of one of the standard calorimeters that circulated through numerous physiology labs and later, in 1913, director of the Kaiser-Wilhelm Institut für Arbeitsphysiologie (occupational physiology), established standard values for proteins, fats, and carbohydrates, which were subsequently refined by Wilbur Olin Atwater (1844–1907), a key figure in calorimeter experiments with humans.[51] Atwater worked with Francis Gano Benedict and physicist Edward B. Rosa (1873–1921) to design a calorimeter that was large enough to accommodate numerous subjects for a period of days, to become a real 'human respiration apparatus'.[52] In addition, the Atwater respiratory chamber welcomed human experiments, including hunger artists. It constituted a seven-foot-long, four-foot-wide, six-foot-high box of copper with a zinc and wood wall, in which humans could eat, drink, work, and sleep, with a constant supply of fresh air, a bed, a chair, and a table. To apply the first law of thermodynamics to human beings, Atwater used the calorie as a unit of measurement and influenced further work of the American Nutrition School, in particular Benedict's experiments[53] (Figure 3.3).

From Ludwig's kymograph to the most sophisticated respiratory calorimeters, hunger artists provided useful cases for experiments on humans in physiology laboratories. From Luciani to Benedict, numerous instruments accompanied the study on Succi, but also on Agostino Levanzin and other hunger artists such as Cetti, Breithaupt, and Merlatti. Again, the sealed boxes of the public shows seemed to be ideal training for the resilience required to stay inside a calorimeter chamber for weeks. In addition, quantified, sound data on professional fasters, obtained using the new instruments, strengthened the experimentation on humans (and hunger artists) in a context of increasing criticism – in particular in the Anglo-American world – of the moral price of animal research.

Figure 3.3. General view of respiration calorimeter by Wilbur O. Atwater and Francis G. Benedict (1903).

Experiments with humans required skilled designs, reliable instruments, and gifted doctors, as well as voluntary laboratory technicians and medical students, to provide additional data for the quantifying dream of experimental physiology.

Partners

Doctor–faster 'pairs' merged the silence of the laboratories with the noise of the public performances, the experimental rigour of the instruments with the slippery behaviour of the hunger artists, the utopian objectivity of the academic elite with the cacophony of charlatan culture.[54] Physiologists often searched for healthy, disciplined, self-controlled adult men (and some women), in a balanced mental state, of standard height and weight, able to establish a trustworthy relationship with the doctor, and capable of regularly undergoing medical tests – fasting in our case – and living in strict confinement for a long time.[55] Nevertheless, hunger artists' reputation for charlatanry and psychic instability (see Chapter 4) made the choices hard. Drs Maréchal and Monin had a smooth relationship with Merlatti in Paris in 1886, but tension arose, as discussed, when the faster wanted to prolong his fast towards the limits of human resistance. Luciani managed to establish a mutually beneficial relationship with Succi in Florence in 1888, but suspicion of insanity always surrounded the faster. Cetti's character as adventurer and balloonist seemed unsuitable for the discipline required in the

Figure 3.4. Stefano Merlatti during his fast in Paris in 1886, reading *Le Voltaire*.

calorimeter chamber, but in the end he gained visibility among the medical profession through being studied by prestigious physiologists in Berlin, alongside Breithaupt.[56] In other cases, things went worse. Benedict, for example, had serious troubles with Agostino Levanzin, and used his own graduate students in some experiments. All of these cases, though, shed light on the subtleties of the experimenter–experimentee relationship and the epistemological status of hunger artists as objects of scientific inquiry.

The partnership between Maréchal/Monin and Merlatti seemed to work well in Paris in 1886 (Figure 3.4). After following Merlatti's daily evolution during his fast in great detail, the two doctors published a book that summarised the unique experience and increased their public reputation.[57] For Maréchal and Monin, who did not belong to the elite of the profession, the book and their study of Merlatti's fast constituted a positive move in all senses, with scant evidence of disagreements with the hunger artist. But tensions between Merlatti and some members of the medical board were frequent. Often putting his health at risk (suffering from acute stomach pain and frequent vomiting), Merlatti disregarded the board's advice to break his fast on medical grounds. Remarks from Merlatti's medical board, as early as his eighth day of fasting, are worth reproducing here: 'Gathered together in the flat he occupies on the first floor of the Grand Hotel, in Paris, and which was graciously made available by the management of said hotel, in the presence of Mr Mario Carl-Rosa, director of the Champs Elysées Academy of Fine Arts, and Mr Ragot, painter; after having read the diary of the experiment and the urine analyses performed by Mr Edmond Vasseur, pharmacist … it was decided that, despite the relatively good condition of Mr Stefano Merlatti, there might be some danger in continuing, for the sake of mere curiosity, an absolute fast which has already lasted eight days without any interruption.'[58] In fact, probably due to the risk of not following the doctors' advice, Merlatti died in 1890, only four years after the Parisian performance, as the press reported in the following terms: 'We remember the faster Merlatti. The unfortunate man died as a result of gastritis contracted during his fasts. He loved fasting too much; that is what killed him. If his fate is to serve a purpose, at least, let it be as an example to Succi.'[59] In that case, doctors seemed incapable of properly supervising Merlatti's decisions. In discussions between the experimenter and the faster, the issues of human resistance to inanition were brought to the fore,[60] often reinforcing doubts about the reliability of the subjects under study. In his experiments with Succi in Florence, Luciani referred to Chossat's work, published in 1843, which established the rate of tissue loss during inanition in animals.[61] Luciani considered Chossat, professor of physiology at Geneva, and Friedrich Tiedemann (1781–1861), professor of anatomy and physiology at Heidelberg, as the key precedents in inanition studies. For more than half a century, Chossat's canonical text of 1843 had been quoted by generations of physiologists.[62]

The contract signed by both Luciani and Succi at the Florence Accademia Medico Fisica for the experimental study of his thirty-day fast in 1888 tried to 'rationalise' their partnership: Succi had the freedom to continue his fast for as long as he wished, but the Accademia and the medical board could stop the study when they considered that his health was at risk, when there was any suspicion that Succi had broken his fast, or if Succi refused to undergo the required tests and experiments. During the fast, Succi had to live in the place designated by the board, and he could only ingest a saline purgative, alkaline water, and still water to wash his stomach. At the end of the fast, if all the rules had been strictly followed, the Accademia would provide Succi with an official certificate of his achievement.[63]

Despite the strict rules of the game, the complexities of Succi's character contributed to Luciani's pledge for a flexible interpretation of the experiment in the framework of a holistic physiology that included physical–chemical but also vital and psychic phenomena, and even some philosophical considerations (see Chapter 4).[64] In that context, Succi's case acquired a new, intriguing dimension. Before fasting, in a personal interview with the doctor, Succi tried to convince Luciani of his reliability as an object of scientific experimentation, stepping back from the danger of going too far, beyond the limits of human resistance to inanition as Merlatti did in Paris, in 1886:

> LUCIANI: I appreciate the originality and simplicity of the means you have employed to make your fortune. With your fasts you have managed to gain a reputation.
>
> SUCCI: But I have not yet managed to make my fortune, dear Professor [...]
>
> LUCIANI: What would you like to conclude? Considering only your more robust constitution, I can easily explain your greater resistance compared to Merlatti. But why didn't you make your superiority more evident by prolonging your fast beyond the terms set by Merlatti?
>
> SUCCI: Do you want to know why? To prove that I am of sound mind and that I'm not at all crazy. I did not want to consume my body for mere stubbornness. My proposal was to fast for thirty days; when the deadline had been reached, I released myself from any obligation.[65]

Although some of Luciani's colleagues in Florence expressed doubts about Succi's reliability as an object of medical experimentation, he moved on intellectually and used Succi's data and his personal acquaintance as a useful tool for his career. Results of the famous experiment in Florence appeared in several of Luciani's textbooks and academic papers, to such an extent that the Luciani–Succi partnership can be considered a key landmark of the experimental physiology of hunger artists.[66]

The scientific study of Cetti (and the almost-unknown faster Breithaupt) was also linked with the relationship with different doctors. In 1887, the press described Francesco Cetti's fast in the following terms: 'Berlin has now started its fasting man, who is called Francesco Cetti. He has commenced a fast which is to last thirty days. Many journalists, physicians and others attended the commencement of this undertaking. Cetti is a sickly-looking man of twenty-seven ... During the fast he is to take nothing but distilled water, or Vichy water. Two physicians will watch him day and night, and two professors have undertaken the scientific experiments and observations.'[67] In fact, Curt Lehmann's and Nathan Zuntz's experiments on Cetti in the first ten days of his fast provided useful data on the composition of his urine, faeces, and carbon dioxide output figures. Some years later, in 1893, a larger team of renowned physiologists, Lehmann,

Friedrich Müller, Immanuel Munk, Hermann Senator, and Zuntz, published a joint paper in the *Archiv für pathologische Anatomie und Physiologie und für klinische Medizin*, the prestigious medical journal edited by Rudolf Virchow. They reported on details of experiments with Cetti and Breithaupt to assess the effect of hunger on nitrogen exchange.[68] But doctors encountered problems when adjusting the experiments to Cetti, since the latter complained about being placed in a cold room, which was very difficult to heat, and about being, in his view, subjected to too strict a surveillance. The fasting room contained several instruments and a sofa on which Cetti lay down for a large part of the day. The sleeping cabinet had a bed and a washstand, and only a table and chairs for the night watch. All doors and windows leading to Cetti's room were locked and sealed, and Cetti's personal belongings were repeatedly examined.[69]

Cetti's personality as a ballooner and traveller did not fit well with the strict rules of the necessary confinement in a small room for days. From his early youth Cetti had led an unsettled life and engaged in various arts and shows to make a living. He had a lively temperament and a certain vanity, especially in his desire to attract the public attention in newspapers. Cetti himself was convinced of the importance of his fast for the advancement of science, but his character was difficult for doctors to manage, the latter expressing some doubt about the genuine value of human experimentation during fasting when compared to animals.[70] In spite of this, doctors sought to convince their audiences that Cetti was a reliable object of experimentation. They presented the faster as a healthy person, who had never been seriously ill. After all the experiments on Cetti and Breithaupt, the physiologists in charge concluded that both fasters shared a strong albumen rate and a relatively high chlorine excretion, the peculiarities of intestinal putrefaction with regard to the formation of aromatic substances (phenol or cresol, indole), the very abundant formation of acetone and acetoacetic acid, and a great quantity of water release. In spite of the trouble taken to adapt Cetti to the quantifying discipline of the laboratory, his body variables during inanition became a key part of the medical study (see Table 3.1).

However, they did not see many differences when comparing the process of fasting in animals and in humans:

> What we have found out about the processes in fasting humans does not differ in principle from the results obtained from animal experiments. Nevertheless, our investigation, even where it only confirms or generalises the findings in animals, does not seem superfluous, if only because in many cases the clinician is only able to correctly interpret his clinical pictures, which are complicated by inanition, through knowledge of the parallel processes in starving healthy humans ... we trust that our studies and those of Luciani, which were carried out only after

Table 3.1 *The quantification of Francesco Cetti's fast by Curt Lehmann and Friedrich Müller*

Day	Body weight (kg)	Water drunk (ml)	Urine excreted (ml)	Decrease in weight (kg)	Red blood cells (per m^3)	White blood cells (per m^3)	Haemoglobin (%)	Pulse (morning)	Pulse (evening)	Temperature (morning, °C)	Temperature (evening, °C)	Remarks
11 March (before fast)	57.080	–	–	–	57200000	normal	115–118	–	–	–	36.2	
12 March (beginning of first day of fast)	57.000	790	1150	–				84				
Fasting day 1	56.450	1200	990	760	–	4800	–	80	72	36.8	36.6	
〃 2	55.510	575	940	575	–	4200	–	76	72	36.6	36.8	Midday walk
〃 3	54.430	1205	1080	1205	–		–	70	72	36.7	36.6	
〃 4	53.450	1500	1310	1170	–		110	68	80	36.4	36.8	
〃 5	52.600	1200	980	1070	5287000		–	80	80	36.7	36.4	Abdominal pain
〃 6	52.350	1625	945	930	–		–	84	80	36.7	36.5	Visit from parents.
〃 7	52.350	1600	995	605	–		–	80	92	37.2	37.4	Abdominal pain.
〃 8	51.840	1200	790	920	–		–	84	78	37.1	37.1	Morning bowel
〃 9	51.240	900	775	548	6830000		85–90	92	87	36.7	36.6	movement (177 g.)
〃 10	50.650	1000	620	970	–		–	80	72	36.2	36.5	Visit from parents
Eating day 1	51.250	–	740	508	–	–	–	80	–	36.5		
〃 2	54.630	–	1507	–	656000	12300						
Normal, 5 April	–	–	–	–	573000	7950						

Source: Curt Lehmann, et al., 'Untersuchungen an zwei hungernden Menschen', *Archiv für pathologische Anatomie und Physiologie und für klinische Medizin*, 131, supplement (1893), 16.

our brief report on the first starvation experiment and which have since been communicated in detail, will form the solid starting point and secure basis for future investigations into the state of starvation and the pathological processes related to it.[71]

Some years later, in 1912, Benedict looked for an ideal hunger artist as a subject for his experiments (Figure 3.5).[72] He was particularly interested in the way in which nutrition, food, exercise, and sleep influenced human and animal metabolism. From 1907 onwards, long journeys to Europe enabled him to contact the leading nutrition laboratories (see Chapter 1);[73] he was later appointed director of the Boston Nutrition Laboratory by the Carnegie Institution of Washington.[74] At that time, Succi was at the end of his career as a hunger artist and had left behind an enormous record of travels and performances, all combined with his habitual role as an object of scientific research in different laboratories. So, in Benedict's view, Succi was an ideal candidate to subject his body to a careful and often uncomfortable set of experiments. But Succi was probably too old at the time, too expensive due to his reputation, and with too many experiments under his belt. In Benedict's view, other candidates suffered from 'nervous diseases', which made them less reliable for experiments. After consulting some of his fellow doctors abroad, he finally chose Agostino Levanzin (1872–1955) from Malta.

In 1911, Levanzin had contacted Benedict expressing his desire to undergo experiments on fasting and forwarding a series of letters of recommendation from several doctors.[75] At seventeen years of age Levanzin had embarked on an ecclesiastical career and later studied medicine for four years. After suffering a severe shock described as 'neurasthenia' in original sources, he married in 1900. Levanzin was a vegetarian, an Esperantist, a man passionate about politics, telepathy and spiritualism, and the founder and later president of the Society of Psychic Studies in Malta (see Chapter 4).[76] Benedict perceived Levanzin as: 'a man of a much higher level of intelligence and intellectual training than Succi. At all times during the fast he was very eager to enter into discussions upon abstract subjects such as the value of the Esperanto language, the political conditions in Malta, the possibility of mental telepathy, and theories of spiritism, as well as the value of fasting. It could not be observed that there was any diminution of his argumentative powers or lack of lucidity of expression. When aroused to counter argumentation he showed the same energy in reply at the end as at the beginning of the fast.'[77]

But, the Benedict–Levanzin experimental partnership also had its dark side. Benedict and his Wesleyan school of fasting were keen to measure the direct determination of heat exchange during inanition. However,

Figure 3.5. Signed portrait of Francis Gano Benedict (n.d.).

since Levanzin had to remain inside the respiration chamber for thirty-one days – the entire fasting period – the experiment became rather impracticable.[78] Levanzin 'complained to the press about the cruel treatment he allegedly received during a … fast in a respiration calorimeter, a room-size device measuring heat loss'.[79] This was an accusation that Benedict never accepted in public, as he personally stated in the *New York Times* in the following terms: 'Dr Benedict … absolutely refuses to make any reply to the charges of brutality made by Professor Agostino Levanzin, who fasted 31 days in the Carnegie Nutrition Laboratory in Roxbury. Levanzin asserted today that he had been squeezed like a lemon and then put out while he was weak and starving.'[80] Three years later, in 1915, in a paper that summarises all the experiments undertaken with Levanzin, Benedict still insisted that: 'The psychological studies and the general observations of the man showed that for thirty-one days the subject was able to exist in a fairly normal mental condition, entirely out of proportion to the physical decline in the body function.'[81]

Tensions between Benedict and Levanzin roused the interest of medical students, who often suffered similar treatment as objects of medical investigation. A Harvard student named M. F. Tuffts described Levanzin as 'a prisoner in a zinc air-tight coffin, with his mouth sealed and breathing air … by tubes in his nostrils', and used the opportunity to also denounce the repulsive way in which dog vivisections were practised at that time.[82] After that incident, Levanzin's reputation seemed to evaporate. In spite of the $400 he received for the experiment, he was later found penniless, begging for food and shelter.[83] Again on the fringes of medical authority, in the 1920s, close to the time of Kafka's short story on hunger artists, Levanzin was fined for the irregular practice of medicine. Paradoxically, though, the press acknowledged that: 'he [knew] more about the care of the body than many young medical graduates will learn in twenty years, if ever'.[84]

In spite of the personal tension, in his publications Benedict acknowledged how useful Levanzin had been for the advancement of the science of nutrition during inanition.[85] Doctors enthusiastically brought hunger artists to their laboratories in the hope of further clarifying the limits of human resistance to prolonged fasting. Others promoted hunger artists as examples reinforcing their eclectic medical positions, at times causing controversy. Long stays in closed rooms or, worse still, in the closed boxes of respiration calorimeters probably did not help. Nevertheless, experimental partners mattered. They became key actors in the experimental culture of the land of the hunger artists when confined in physiology laboratories. Peculiar biographies, blurred boundaries of expertise, and, sometimes, personal disagreements made things harder but, on the whole, the partners became a key pillar of the science of inanition in doctors' attempts to 'isolate' their patients from the public commotion of these controversial performances.

Metabolism

In 1915, Benedict pointed out that: 'Not infrequently the observations made in these professional fasting exhibitions have contributed materially to the sum of human knowledge, since there is an intense physiological interest in the *vital processes during such prolonged abstinence from food*.'[86] The assertion was consistent with a longstanding interest within physiology to understand the reaction of the human body during processes of inanition as well as in the re-feeding phases at the end of the fasts. In the 1960s and 1970s, Benedict's work was re-discovered by several schools of physiology that conceived inanition and starvation as a series of different metabolic phases.[87] It was probably a consequence of doctors' earlier belief, in the time of the hunger artists, that inanition was the simplest form of nutrition and a key test to study human metabolism, that is to say, the chemical processes that allow life and the normal functioning of the body. Therefore, from a historical perspective, the professional fasters (Tanner, Succi, Jacques, Merlatti, and Levanzin, among others) procured the key quantitative data for the study of human inanition and contributed to the early experimental studies on metabolism.[88]

In 1907, Benedict noted that, since metabolism was considered as a twofold problem – *catabolism* as the process of disintegration of the body material, and *anabolism* as the transformation of food materials into body materials – 'by withholding food, the anabolic activities may be depressed to such an extent as to make the study essentially one of catabolism. Consequently, *studies of transformations in the body during inanition are of great value and, logically, should precede the studies in which anabolic and catabolic processes are combined.*'[89] Some years later, in 1915, Benedict again praised the scientific value of experimental studies on human fasting and stressed the link between diseases and the stages of nutrition. Since many illnesses 'border on complete or nearly complete inanition', he considered that the physiology of fasting was 'of special importance to the physiologist and the clinician alike'.[90]

In fact, the absence of food intake did not interrupt metabolism, since the body survived on material previously stored in tissues.[91] Carbon, nitrogen, oxygen, and hydrogen entered animals through food ingestion and were later excreted, being part of the nutrients exothermically burned through respiration. Animal heat was the result of the combustion of carbohydrates and fats releasing carbon dioxide, whereas motion was caused by the decomposition of proteins into urea and uric acid. In 1847, Hermann von Helmholtz's early formulation of the first law of thermodynamics was inspired by the idea that animals consume a certain amount of chemical energy to produce heat and work.[92] Some decades later, the three main phases of human metabolism were understood to comprise the conversion of food intake into energy, into building blocks for cells and tissues, and into waste for elimination; in their inanition experiments, hunger artists had input into this understanding around 1900.

Back in the time of the hunger artists, Succi's case, in particular, attracted the interest of many physicians and provided new, valuable data. After Luciani's pioneering work in Florence in 1888, prestigious physiologists published papers in reputed journals of physiology on the scientific study of Succi's fast and its value for further research on human metabolism,[93] as well as in the cases of Cetti and Breithaupt already discussed.[94] A decade later, in 1898, Atwater and Charles Langworthy made extensive use of hunger artists in their compilation of experiments on metabolism.[95] Aiming to quantify the balance of carbon and nitrogen, they included sixty-two tests with fasting men and four with fasting women, which included well-known professional fasters, with partial or complete inanition. They also tested diseased subjects consuming no food, and fasting animals. All this research was carefully reported and summarised in detail under the heading: 'Experiments in Which the Subjects Were Fasting'.[96] Tables included the date of publication of the experiment, the experimenter's name (the observer), details of the subject of experimentation (occupation – many of them being professional fasters – age, weight, food intake if any per day, duration of the fast, nitrogen exchange in food, urine, and faeces), plus some particular remarks about every case. The list of the subjects of experimentation included categories such as man, woman, insane man, professional faster, soldier, and laboratory servant.[97]

In the early twentieth century, Edward Cathcart (1877–1954), professor of physiological chemistry at the University of Glasgow and later research associate in the Nutrition Laboratory at the Carnegie Institution in Washington, studied the metabolism of hunger artist Victor Beauté.[98] In Cathcart's view, this kind of experiment provided useful information about the catabolism of protein through the products excreted in urine. After fasting for fourteen days, Beauté was put on a diet consisting of cream and starch, and within three days the output of nitrogen had fallen to a third of the output on the last day of the starvation period.[99] In 1909, Dr Graham Lusk (1866–1932), another pupil of von Voit and at that time professor of physiology at the Cornell University Medical College in New York, summarised in his *Elements of the Science of Nutrition* (1909)[100] the quantification of Cetti's proteins, fat, and calories during his fast (see Table 3.2).[101]

Lusk devoted a chapter to the metabolism of starvation, with special emphasis on the measurement of nitrogen in urine as an indicator of protein decrease, in particular referring again to experiments on hunger artist Cetti by Lehmann and Zuntz; he compared daily nitrogen excretion in different fasts by Cetti, Breithaupt, a medical student, and Succi (see Table 3.3).[102] In the weight loss during starvation, Succi's data were also compared with those of animals such as dogs.[103] Lusk concluded that: 'the minimal metabolism requirement of the fasting organism appears remarkably constant in different men ... During prolonged fasting the nitrogen output sinks much below the figures of the early days.' Lusk also pointed out that, if the organism had been previously well nourished, the fasting metabolism was remarkable even among all hunger artists, about 13 per cent of the total energy being derived from protein and 87 per cent from fat.

Lusk's comparative efforts among hunger artists included detailed urinary analysis presented in tables in his *Elements of the Science of Nutrition*, such as the cases of Succi and Beauté (see Table 3.4).

Table 3.2 *Metabolism of Francesco Cetti in starvation*

Fasting days	Protein (g oxidised)	Fat (g oxidised)	Calories from protein	Calories from fat	Calories total	Calories per kg of body weight
1–4	85.88	136.72	329.8	1288.2	1618	29.00
5–6	69.58	131.30	267.3	1237.4	1504	28.38
7–8	66.30	149.35	254.7	1407.3	1662	31.74
9–10	67.96	132.38	261.1	1247.4	1508	29.26

Note: Graham Lusk's quantification of Cetti's proteins, fat, and calories during his fast.
Source: Graham Lusk, *Elements of the Science of Nutrition* (Philadelphia: W. B. Saunders, Co., 1906), 63.

Table 3.3 *Nitrogen excretion (g) in different hunger artists during the first days of fasting*

Day	Cetti (1)	Breithaupt (2)	Succi (3)	J. A. (4)	Succi (5)
1	13.55	10.01	13.811	12.17	17.00
2	12.59	9.92	11.03	12.85	11.20
3	13.12	13.29	13.86	13.611	10.55
4	12.39	12.78	12.80	13.69	10.80
5	10.70	10.95	12.84	11.47	11.19
6	10.10	9.88	10.12	–	11.01

Note: 'J.A.' refers to an unknown faster.
Source: Graham Lusk, *Elements of the Science of Nutrition* (Philadelphia: W. B. Saunders, Co., 1906), 66. Lusk compiled the table using the following sources: (1) Munck, *Archiv für Pathologische Anatomie und Physiologie und für klinische Medizin*, 13 (1893), 131, Supp., 25; (2) ibid., 68; (3) Luigi Luciani, *Das Hungern. Studien und Experimente am Menschen* (Hamburg and Leipzig: Leopold Voss, 1890); (4) J. E. Johansson, E. Landergren, K. Sonden, and R. Tigerstedt, *Skandinavian Archiv für Physiologie*, 7 (1896), 54; (5) E. Freund and O. Freund, *Wiener klinische Rundschau*, 15 (1901), 91.

Lusk also referred to Benedict's pioneering experiments on the metabolism of glycogen – the store of carbohydrates in bodily tissue – in inanition, in particular Benedict's measurement of the metabolism of glycogen in seventeen experiments on seven volunteers, using a respiration calorimeter. During seven days of fasting, Benedict quantified the loss of proteins, fats, and glycogen, as well as the calories released and the content of nitrogen in urine. Lusk's description of Benedict's detailed procedure is worth mentioning in detail here:

Benedict's fasting individuals were placed in a respiratory calorimeter, and in addition to the usual routine the amount of oxygen consumed by them was measured. Knowing the last factor Benedict was able to calculate the amount

Table 3.4 *A comparative urine analysis of Giovanni Succi and Victor Beauté during inanition*

(a) *Complete urinary analysis of Succi on first, third, eleventh, and twenty-first days of fasting (weight in grams)*

	Day of fasting			
	1st	3rd	11th	21st
Urine, cc	1435	575	378	235
Total N, g	17.0	10.55	6.32	2.82
Urea N, g	14.8	9.65	5.64	1.65
Uric acid N	0.29	0.20	0.075	0.046
Purin base N	0.13	0.064	0.042	0.034
Creatinin N	0.134	0.198	0.372	0.025
Ammonia N	0.43	0.144	–	0.10
Total S	3.2	1.3	0.8	–
Total P_2O_5	2.98	2.52	0.41	0.64
Cl	14.9	2.56	1.51	0.7
Ca	0.25	–	0.31	–
Mg	0.33	–	–	–

Source: Graham Lusk, *Elements of the Science of Nutrition* (Philadelphia: W. B. Saunders, Co., 1906), 68.

(b) *Urinary analysis of Beauté on the first, third, twelfth, and fourteenth days of fasting (weight in grams)*

	Day of fasting			
	1st	3rd	12th	14th
Total N	10.51	13.72	8.77	7.78
Urea N	8.96	12.26	6.62	5.99
Ammonia N	0.40	0.73	1.05	0.73
Uric acid N	0.12	0.06	0.17	0.17
Purin base N	0.029	0.032	0.023	–
Creatinin N	0.42	0.34	0.30	0.24
Creatin N	0.02	0.09	0.09	0.10
Total S	0.614	0.801	0.577	0.536
Total P_2O_5	2.26	2.98	1.55	1.25
Cl	3.2	1.5	0.18	0.24
Ca	–	0.216	–	0.096
Mg	–	0.131	–	0.037
K	–	1.33	–	0.515
Na	–	0.865	–	0.096

Source: Graham Lusk, *Elements of the Science of Nutrition* (Philadelphia: W. B. Saunders, Co., 1906), 69.

of nitrogen destroyed by deducting from the total oxygen intake the part necessary to oxidize the protein catabolized, and then, in the light of the knowledge of the respiratory quotient apportioning the remainder of the oxygen to the non-protein carbon dioxide eliminated, in such a way as to indicate the amounts of the glycogen fat destroyed. The heat value of the metabolism thus calculated agreed with the heat as actually measured by the calorimeter in which the man lived within one half of one per cent, whereas if the non-protein carbon of the first day had been reckoned as fat metabolized, as had heretofore been the custom, the discrepancy would have been as high as five per cent in some instances. This shows the usefulness of a comparison of direct and indirect calorimetry.[104]

In 1915, Benedict published a key paper that summarised in a graph all the measurements his team took three years earlier for the study of Levanzin's metabolism during fasting (Figure 3.6).[105] In order to draw a metabolism chart, he quantified several variables for the thirty-one days of the fast: oxygen and carbon dioxide, alveolar CO_2 tension, blood pressure, heat per 24 hrs, body temperature, heat per kilo per hour, respiratory quotient, respiratory rate, pulse, chlorine, total nitrogen, phosphorus (P_2O_5), carbon in urine, oxobutyric acid, uric acid (nitrogen), total sulphur, and ammonia (nitrogen). The study also covered other aspects such as an assessment of the intestinal flora and skin excretion, a complete chemical examination of urine (including nitrogen, chlorine, phosphorus, sulphur, acidity and B-oxobutyric acid), a determination of the mineral constituents (calcium, magnesium, sodium, potassium), and obviously the intake–output balance of matter and heat through the respiration calorimeter.[106]

Some years later, Sergius Morgulis (1885–1971), professor of biochemistry at the University of Nebraska College of Medicine, and another of the key names in nutrition textbooks from the early twentieth century, compared the vital signs of different hunger artists and explored the metabolism of the limits of life.[107] In 1923, just a year after the appearance of Kafka's short story, Morgulis published a comprehensive book titled *Fasting and Undernutrition: A Biological and Sociological Study of Inanition.*[108] Under the banner of the law of conservation of energy applied to the human body, Morgulis described four different periods in any process of prolonged fasting, which refined and complemented Luciani's earlier attempts. Each period represented an approximate loss of 1/8 of the body weight. In the first period glycogen stores decrease rapidly and the ability to oxidise carbohydrates diminishes. In the second and third stages, energy is mainly produced from stored fat and the oxidation of glucose reduces to a minimum. The fourth phase leads to the critical exhaustion of body fat and a critical scarcity of proteins.[109] Referring to numerous

METABOLISM CHART OF A MAN FASTING FOR 31 DAYS

14 APRIL–15 MAY 1912

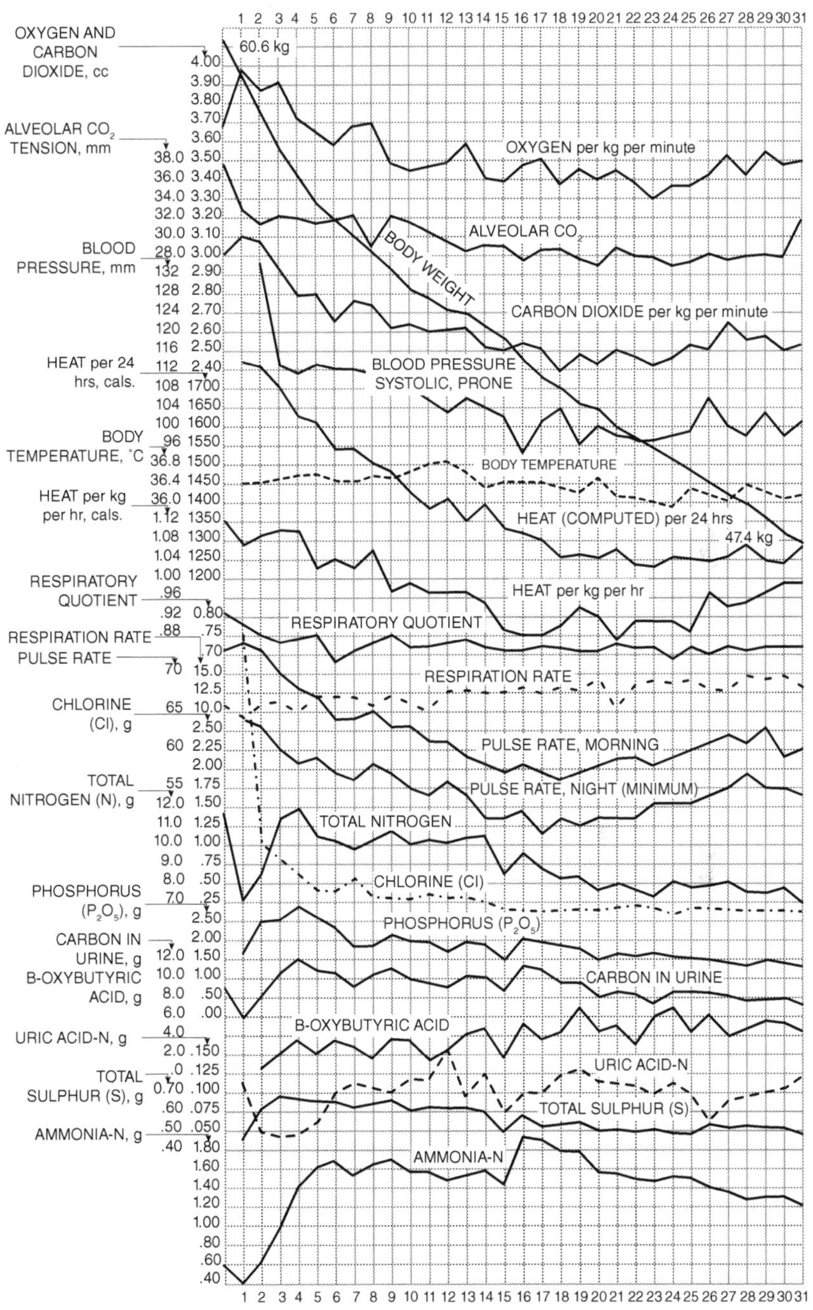

Figure 3.6. Francis Gano Benedict's graph of Agostino Levanzin's metabolism during his thirty-one days of fasting in 1912.

Table 3.5 *The measurement of daily nitrogen excretion in hunger artists during the first four days of fasting (g)*

		Day of fasting			
Subject	Normal	1st	2nd	3rd	4th
Beauté	16.45	10.51	14.38	13.72	13.72
Succi	17.85	15.19	12.13	15.25	14.08
Tosca	13.99	8.76	8.38	10.73	9.40
Succi	8.99	8.72	8.45	9.05	8.51
Levanzin	11.54	7.10	8.40	11.34	11.87
Kazawa	12.05	10.90	13.73	12.91	15.11

Source: Sergius Morgulis, *Fasting and Undernutrition: A Biological and Sociological Study of Inanition* (New York: Dutton and Company, 1923), 153.

experiments on hunger artists, Morgulis stressed that they were particularly useful to measure the time that a normal body state could be preserved to study the varied resistance in different organs and organisms, and finally, again, to understand the intriguing process that leads to human death.[110]

In discussions on the quantitative measurement of daily nitrogen excretion, Morgulis again used data from hunger artists Beauté, Succi, the fasting woman Tosca, Levanzin, and Kozawa (see Table 3.5). In other cases, to measure weight loss during the inanition process, Morgulis added data from Succi's successive fasts (see Table 3.6). Morgulis quantified the changes in weight and the speed of loss in body weight, changes in the composition of the organism, and metabolism of matter and energy; that is to say, he made a careful study of the chemical processes that occur within a living organism during prolonged fasting.[111]

From Luciani's early approaches to the physiology of inanition to the more explicit approaches to human metabolism by renowned doctors such as Atwater, Langworthy, Cathcart, Lusk, and Morgulis, there is considerable evidence of the key role played by hunger artists as experimental objects for the study of human metabolism under conditions of food deprivation. There is no doubt that prolonged fasts provided an excellent opportunity to study protein delay during catabolism, but also to gain a better understanding of the different adjustments of the body – its capacity to store energy and control its allocation – during inanition and re-feeding.[112]

***

Table 3.6 *The quantification of weight loss of different hunger artists during their fasts*

Subject	Kozawa	Beauté	Schenk	Succi							Jacques	Levanzin	
Initial Weight (kg)	50.7	65.8	56.4	63.5	65.2	71.7	63.5	63.0	61.0	55.9	62.3	60.5	Average loss
Duration of fast													
14 days	13.3	11.9	13.3	12.9	10.6	11.1	13.3	15.7	12.6	12.7	11.3	12.4	12.59
16 ″			14.4	13.1	11.2	12.1	14.2	16.3	13.8	13.8	11.8	13.8	13.45
20 ″				16.7	13.2	13.7	16.6	18.1	16.6	16.1	13.0	16.0	15.67
29 ″							20.3	21.8	20.9	20.6	16.4	20.7	20.11
30 ″								22.6	21.4	20.8	16.6	21.4	20.56
31 ″										21.0		21.9	21.45
40 ″										25.3			25.30

Note: Succi's fasts took place at Naples, Rome, Zurich, Florence, Paris, Milan, and London respectively.
Source: Sergius Morgulis, *Fasting and Undernutrition: A Biological and Sociological Study of Inanition* (New York: Dutton and Company, 1923), 90.

Beyond the knowledge already acquired from experiments on animal starvation, which was never free of controversy and moral doubts, professional fasters provided new data for the study of human inanition. They became useful agents to test and legitimise the new instruments, from the kymograph to the sophisticated respiration calorimeters. In their negotiations with doctors, they acted as relevant historical actors who reshaped the experimental culture of the physiology of inanition. Moreover, hunger artists also became valuable objects of experimentation for the clarification of several aspects of human metabolism: the rate of tissue decrease during inanition, the body's overall resistance to the continuous lack of food, and details of the process of human starvation and final death. The latter was a standard procedure with animals, but now hunger artists opened up a new, more nuanced, fruitful path of research. Maréchal, Monin, Luciani, Lehmann, Müller, Benedict, Cathcart, Lusk, Morgulis, and a long list of doctors did their best to bring hunger artists to their laboratories for 'scientific' study, and to 'isolate' them from the commotion and cacophony of the public sphere. However, this appeared to be an impossible endeavour and, to some extent, an epistemological contradiction as far as the genuine culture of the faster was concerned, always falling somewhere between science and hype.

As in other experimental systems, the physiological study of hunger artists in laboratories had its own epistemic, technical, social, and institutional limitations, with no clear boundaries between the four realms. In spite of the numerous optimistic claims that appear in the former sections, there was no full consensus on the sound additional knowledge that experiments with humans could provide when compared with the data already accumulated on starving animals. Equally, in spite of the initial quantifying euphoria of the new instruments of physiology and the impressive collecting endeavour of the original design and evolution of the respiration calorimeters, non-quantified factors seemed to remain in the physiology of prolonged fasting. Faculties of medicine, research schools of physiology, prestigious textbooks, and academic papers substantially contributed to the knowledge of hunger and fasting, but again public controversies and endless commotion partly weakened the doctors' over-optimistic claims of objectivity and quantification in the laboratories.

In 1890, at the height of the public success of hunger artists, the *British Medical Journal* categorically stated that: 'Experiments on public performers are never satisfactory.'[113] Thirty years later, Morgulis, one of the American doctors who had worked with an enormous amount of experimental data on hunger artists in his studies on human metabolism and nutrition, insisted that: 'the therapeutic value of inanition … should be studied experimentally and not be left to the judgment of amateur enthusiasts'.[114] This experimental euphoria fitted very well in the context of new experimental physiology and its quantitative instruments, as well as in the doctors' expertise. Nevertheless, in a medical culture that faced plural views and eclectic, heterodox practices, and experienced the emergence of new medical specialities, doctors' attempts at objectivity seemed an impossible enterprise.

As Luciani's case makes clear, the sophisticated quantitative experiments and the complex relationship with the object of experimentation did not necessarily lead to a strict materialist, quantitative way of understanding the human body in a process of prolonged fasting close to starvation. For a better understanding of the limitations of doctors' experimental euphoria, we shall again open the doors of the laboratories and look at other places, practices, and disciplines that also appropriated hunger artists as useful objects of study.

4 Spirits

Although an emphasis on 'objective' measurements was a general trend in the second half of the nineteenth century,[1] physiology was never 'unified, neither theoretically nor practically'.[2] Instruments, experiments, graphs, numbers, and tables coexisted within a context of blurred medical boundaries. Experiments on hunger artists within the laboratory walls of academic, 'orthodox' institutions usually clashed with contemporary trends of 'heterodoxy'.[3] This was not incompatible, though, with frequent calls for a 'scientific' study of practices such as spiritism, mediumship, ghosts, and other paranormal phenomena such as *tables tournantes*, *somnambules*, *fantômes*, and ectoplasmas. They attracted the interest of renowned scientists such as Alfred Russell Wallace, William Crookes, and Oliver Lodge, and well-known popularisers such as Camille Flammarion.[4] Pierre and Marie Curie addressed mediumship; Flammarion combined his popular astronomy with the practice of spiritism; Charles Richet invented a new and controversial discipline, 'metapsychics', in order to accommodate the study of diverse paranormal phenomena.[5] As described by a contemporary witness, in 1887: 'in the century of positivism everything supernatural was denied, rationalism sought to explain everything with naturalism, and never, since the Middle Ages, has the supernatural ever been so affirmed in the eyes of all, at the doors of all'.[6]

Nevertheless, in the 1930s, the 'occult' sciences of the times of our professional fasters became 'pseudo-sciences' and were progressively marginalised away from the interests of academia.[7] Hunger artists seem, therefore, closely tied chronologically to the golden years of occult practices. They took advantage of that atmosphere of heterodox phenomena for the sake of their own professional careers, became active driving forces for a plural, eclectic medicine often dealing with opposed world views, and opened the door to varied interpretations of the supernatural. Those times of heterodoxy also coincided with contemporary debates on the aims and status of physiology as an academic discipline. Scientific optimism and physico-chemical reductionism often coexisted with a holistic conception of physiology that could potentially integrate qualitative, vitalist, even 'psychological' explanations of the functioning of the human body. However, the emergence of new medical specialities made things even more puzzling[8] and opened new spaces for a psychological turn.[9] From the 1860s onwards, the emergence of psychology as an independent discipline was the result of a complex intersection of different traditions: philosophy (and natural philosophy), physiology, and early sociological approaches. It provided an ambiguous territory in which hunger artists and their appropriations could find their own place.[10]

In 1880, Henry Tanner had placed himself on the fringes of official medicine.[11] As described by Dr Robert Alexander Gunn, his biographer, Tanner's interest in fasting could not be disentangled from his controversial medical positions. Gunn explained that 'He [Tanner] had intended entering an allopathic college in the Fall of 1857, but was so impressed by the intolerance of the old-school societies, in their efforts to secure medical legislation at that time against the so-called "irregular physicians", that he changed his plans, and entered the Eclectic Medical Institute of Cincinnati, Ohio.'[12] During his fast, doctors of the New York Neurological Society (NYNS) disputed Tanner's credibility and questioned the team of overseers, since not all of them had professional medical careers (see Chapter 2). That controversy reflected the contemporary medical pluralism in the United States. In the 1880s, beyond the reductionist materialist explanations coming from Europe, and particularly from the most prominent German schools of physiology, Tanner wanted to prove that his fast was healthy, therapeutic and a good example of the power of human will, beyond the restrictions of allopathic medicine.[13]

In Europe, materialism caused controversy in different schools of physiology and steered a huge range of vitalistic, spiritualist reactions against it.[14] The materialist school of Johannes Müller, with pupils such as Émile Du Bois-Reymond, Hermann von Helmholtz, and Ernst Wilhelm von Brucke, alongside Carl Ludwig and the Leipzig Physiological Institute, developed a physiology mainly reduced to physics and chemistry and based on quantitative approaches using the new scientific instruments (see Chapter 3). Helmholtz compared a living organism (and a human being) with a steam engine, both as machines which consume food (and fuel) and produce mechanical work and heat. In this kind of explanation there was no room for vital causes.[15] In a similar vein, Du Bois-Reymond defended the idea that any scientific explanation of a natural phenomenon had to be reduced to motions of atoms and, in tune with Müller's research school, perceived the second law of thermodynamics as an ideal tool to fight against 'irregular vital and mental causes'.[16] Nevertheless, in opposition to these materialist trends, the French physiological tradition, from François Magendie to Claude Bernard and Charles Richet, never fully abandoned vitalistic explanations of the function of the body. They defended a physiology that could not be strictly reduced to physics and chemistry, but could integrate a sort of vital principle with a link to the French vitalistic medical tradition of the eighteenth century.[17] Should the living realm have some degree of autonomy from the laws of physics? If so, there was perhaps room to introduce, for example, Succi's willpower in the explanation of his resistance to inanition. Magendie's heritage had brought to the fore animal experimentation, studies on digestion, glycogen,[18] and carbohydrate metabolism –in particular the role of the liver as a source of sugar – but using few instruments and endorsing qualitative explanations with vitalistic remains.[19]

Moleschott's arrival in Rome in 1879 was a landmark for the introduction of a materialist physiology in Italy. Moleschott never dismissed, however, the more philosophical aspects of the discipline. In fact, one of his pupils, Angelo Mosso, worked in Florence with Moritz Schiff, who came from the Parisian school of Magendie. Having also trained with Ludwig in Leipzig – one of the key figures in the materialistic approach – Mosso designed sophisticated instruments for the quantification of emotions, but always left some room for qualitative, unmeasurable, philosophical ideas, which many Florentine doctors, including Luigi Luciani, considered compatible with their conception of physiology. In 1893, Luciani became Moleschott's successor in the position of chair in human physiology in Rome.[20]

Succi's case fitted therefore very well with Luciani's polymorphic, polyvalent, and even holistic conception of physiology[21] and with his reluctance to accept a materialistic reductionism. Luciani's major conclusion after the experiment with Succi in Florence in 1888 was that, during the fast, the dynamics of matter exchange was a fundamental function of the nervous system, considering the latter as an integrated unity.[22] However, the central, albeit vaguely defined role of the nervous system in the regulation of the body led to alternative, complementary explanations for Succi's resistance. Also, in that framework, the cerebralist–visceralist debate added complementary arguments around hunger and inanition. Cerebralism looked for lesions in the brain as causes of psychic illness, whereas visceralism considered that psychic disorders had to be located in the abdominal viscera. In the second half of the nineteenth century, including our period of study for the hunger artists, cerebralists also placed issues of appetite, digestion, and eating in the realm of the nervous system.[23]

Anti-materialists distanced themselves from the ideological consequences of a doctrine that they often perceived as potentially leading to atheism and social anarchy. In *Le lendemain de la mort* (1872), science populariser Louis Figuier described the vitalist tradition of the Montpellier medical school as a contemporary tool against materialism, which he considered as a danger for the moral order of the Christian tradition.[24] In that sense, anti-materialistic crusades – such as that of the famous populariser of astronomy, Camille Flammarion, who published his *Astronomie populaire* in 1880, the same year as Tanner's seminal fast in New York – seemed compatible with the interest in more 'spiritual' explanations of the natural phenomena.

Hunger artists' performances were ideal examples for those reluctant to accept a strictly materialist explanation of the functioning of the human body and therefore contributed towards strengthening eclectic, plural, and controversial fields and occult forces. They piqued the interest of a number of well-known scientists who moved towards spiritual and psychological explanations of the causes of resistances to inanition. As described in the sections that follow, hunger artists' performances – their public dimension and their widely accepted scientific interest – revealed the shortcomings of materialism and had an impact on heterodox, controversial practices such as spiritism. They also contributed to shaping the contemporary psychological turn, which embraced a gender perspective.

Mediums

While *spiritualism* spread mainly within the United States and Britain, *spiritism* grew strongly in southern European countries. Both spiritualists and spiritists believed in disembodied spirits and their capacity to communicate with living humans. Spiritists believed in the spread of mediumistic activities, but also in reincarnation and the plurality of inhabited worlds, while spiritualists may not.[25] Far from any occult, hermetic approach, spiritism was perceived by many contemporaries as a 'rational' practice submitted to the audiences' view and to empirical demonstration.[26] Moreover, in the 1880s[27] spiritism was becoming increasingly recognised. Prestigious scientists wanted to investigate these phenomena rationally. Female mediums gained a popularity that resulted in feminist demands. The strong ideological charge of the movement represented a potential danger for the hegemony of traditional religion. Allan Kardec's *Livre des esprits* (1857) became canonical reading.[28]

Between his fasts, Succi regularly gave an itinerant public lecture. Titled 'The Life of a Man: The Human Soul and Its Movements', it included Succi's autobiography along with some general explanations and experiments of telepathy, hypnotism, magnetism, and spiritism.[29] There is no historical evidence of audience figures, but suspicion about

Succi's mental health was common at the time. In 1885, Dr Bruno Battaglia examined Succi in Cairo (after his African trips) and concluded that he had made up his account of the way in which he acquired his spiritist faith. Battaglia stated that, in a sort of delirium of the apotheosis of the ego, Succi believed that he had 'died' symbolically,[30] but lived as a lion spirit, that is to say, he became the living dead, immortal and omnipotent. That fictional narrative was part of Succi's strategy to gain social recognition and the starting point of his long professional career as a hunger artist. In 1888, in Florence, when subjected to scientific study, Luciani described Succi's trips to Africa as a key experience for his spiritism (see Chapter 1).[31] Succi apparently resisted liver disease through fasting, and Luciani believed that 'a new spirit had penetrated and acted in him, able to sustain his forces to reinforce his matter. He referred to a spirit combined with immortal matter', which he talked about when he returned from Rome, but that only the spiritist circles believed: 'in their nocturnal sessions they received communications that a stalwart, called Spirito Leone, had arrived in Rome … initials GS.'[32]

Succi's fictional narrative appeared in the daily press and in spiritist journals. In 1902, the French spiritist journal *L'Écho du merveilleux* again brought to the fore the African roots of Succi's spiritism. *L'Écho* described how, in 1879 in Dar es Salaam, Succi encountered an invisible force who spoke to him.[33] *La vie posthume* adapted the same narrative of the African adventures, this time introducing the character of a Zanzibar 'native' who supposedly cured Succi of a strange illness with some herbs. The native was killed, Succi took the herbs, and back in Rome he invoked his soul in a spiritist session in which witnesses reported that Succi himself was in a state of amazing excitement. The man's spirit had advised Succi to keep the herbs, since one day they would be used in a revolutionary discovery that would end the hunger of mankind.[34] In spite of considerable reluctance and widespread suspicion of fraud and insanity, the fictional narrative of Succi's spiritist experience – in contact with a dead native or converted into a lion – gained supporters. Although he never described him as insane, Luciani was convinced that Succi's spiritist faith and his supposed resistance to inanition were the direct result of his delirium.[35] Luciani himself interviewed Succi before starting the fast in his laboratory, and again the spiritist issue came to the fore:

LUCIANI: And would that be? … Spiritism? … Come on, dear Succi, let's speak in confidence … Does it seem serious to you that a man like you, who has travelled a lot and who shows so much judgement, loses himself in such superstition?

SUCCI: I do not claim to convert you to spiritism … on the contrary, if you wish to leave this topic aside, I would be grateful.

LUCIANI: Forgive me, but was Merlatti, who fasted for 40 days at the same time as you, a spiritist too?

SUCCI: And because of that what an impression he made! Didn't you know that his case was abandoned by the medical professionals, who refused to take responsibility for the continuation of his fast? In the last days he was reduced to such extremes that to keep him warm they had to wrap him in cotton wool! That was not an experiment, it was a genuine suicide attempt that the authorities should have prevented [...]

LUCIANI: I cannot blame you. But in the meantime, you too must agree that you can fast not only for 30, but also for 40 days, without the help of any spirit, not even the spirit of wine.

SUCCI: You wish to joke ... Please, let's not return to this topic.[36]

In Florence, Dr Angiolo Filippi considered that Succi's spiritism was simply part of his strategy to draw the attention of the general public to his fast, particularly targeting the most ignorant.[37] Spiritism was an ambiguous terrain in which Succi played his cards in between fictional trips, after-death experiences, and public fasts. In the slippery zones going from academic science to popular hype, he appropriated the contemporary culture of spiritism to strengthen his public reputation and his professional profile as a hunger artist. Whereas academic doctors used hunger artists' performances and controversial status for their own professional interests and philosophical, ideological world views, here, at least in Succi's case, the elusive and blurred frontiers of spiritist practices became useful tools to present hunger artists as respectable professionals.

Succi edited his own journal, *Il Corriere Spiritico*, in Florence in 1888 (Figure 4.1).[38] He vaguely defined his spiritual power as 'an omnipotent force, which materialists deny',[39] and provided extra arguments for the anti-materialist crusade of the time. Succi used his journal to introduce himself as another 'marginalised' genius – like Fulton, Jenner, Galvani, and Mesmer – a victim of indifference, always suspected of charlatanism and fraud. Details of that 'omnipotent force' remained elusive, however, often presented with bombastic words: 'And when our Director Giovanni Succi here in Florence has undertaken and completed his fast, we will make known the method and application of the spirit force by which this grandiose enterprise can be achieved, one of the strangest and most indisputable discoveries, destined to bring about a *revolution in current physiological and psychological science and social well-being*.'[40] Succi served as director of the journal in collaboration with Achille Ricci, the manager. It assembled miscellaneous news, in Italy and abroad, on spiritist experiences, photographs, turning tables, reports from different spiritist societies, publications, books and journal articles, public lectures, translations, and Succi's '*sentenze*' – short thoughts on the nature of the doctrine – such as: 'What is life? For materialists it is a dream that passes

Figure 4.1. Front page of *Il Corriere Spiritico*, 1 (1888).

and never returns, but for those who know what it is, life is real because it is immortal'; 'What is death? Sublime lance that stops those who rise too high'; 'It is not man, that is to say human being, the matter that covers man; the true being is what is enclosed within the Mask of man.'[41]

Succi's spiritism was in tune with the international success of that doctrine. In Italy, 'scientific experiments' with mediums were common and well known. This was the case of the experiment in Milan in 1892, for example, in which renowned luminaries of science such as Charles Richet, philosopher of the occult Carl du Prel,[42] criminologist Cesare Lombroso, sculptor Tommaso de Amicis, and astronomer Giovanni Schiaparelli carefully studied and personally attended a séance with the famous medium Eusapia Palladino.[43] From the 1860s onwards, journals such as *Annali dello Spiritismo* and *Spiritismo* gained wide readership throughout Italy.[44] In France, Flammarion, a good friend of Allan Kardec, maintained that spirits, magnetism, and somnambulism were not supernatural phenomena but had to be scientifically studied. Succi reported on Flammarion's thought in *Il Corriere Spiritico* in the following terms: 'The supernatural does not exist. The manifestations attenuated via the work of mediums, such as those of magnetism and somnambulism, are in the natural order of things and must be rigorously subjected to the control of experiences. There are no more miracles … but the positive study of this new psychology.'[45] In 1887, just a year after the famous Succi–Merlatti contest in Paris, bacteriologist Paul Gibier, the future founder of the New York Pasteur Institute, published his book *Le spiritism* in defence of using the experimental method for the analysis of spiritism.[46]

In 1888, during the Barcelona International Exhibition,[47] after being welcomed by the city mayor, Succi attended the Congreso Internacional de Espiritismo,[48] a conference that mainly drew followers of Allan Kardec. Succi introduced himself as the editor of his journal and attended the conference as a true practitioner of spiritism. At the same time, he sent a short note to the Barcelona Congreso de Ciencias Médicas, seeking medical doctors who were willing to supervise his thirty-day fast and, as had recently happened in Milan and Florence, to become members of his scientific committee.[49] In Madrid, Succi's fast in 1888 also took place under the banner of the spiritist culture of the city. One of the members of the medical board, Dr Anastasio García López, was a committed spiritist as well as president of the Sociedad Espiritista Española, which had been founded in 1872. He also contributed to the edition of the journal *El Criterio Espiritista* that had appeared in 1868, just after a liberal revolution which provided new spaces for freedom of thought and speech. The journal included doctrinaire articles, translations, bibliographies, and descriptions of detailed communications with spirits.[50] As early as 1868, in the first issue of the new journal, Kardec addressed a letter to the founder of the new Spanish society, Enrique Pastor (known as Alverico Perón), which appeared in *El Criterio Espiritista* and again presented the new doctrine as a useful tool to fight materialism: 'Let us hope that the humanitarian and regenerative doctrine of spiritism will not find unbearable official obstacles, and that its supporters will not be persecuted for being dedicated to its dissemination. They will no doubt encounter the obstacle of disbelief, but this opposition is not to be feared, because day by day it grows weaker, and materialism does not have roots in public opinion.'[51]

Spiritist practices were not exempt from fraud, so those who wanted to study it scientifically and fight against deceivers had a lot in common with medical committees and overseers of hunger artists' public performances. In that sense, mediums and hunger artists shared a similar status, as mediators between the elite and popular culture and as objects of challenging and controversial scientific investigations. Although the pro-spiritism profile seems evident in Succi's case, other professional fasters also placed themselves in ambiguous territories. Tanner never renounced the mystical part of his eclectic philosophy, even in frequent confrontations with contemporary medical authorities.[52] Cetti was a juggler and magician for a good part of his life. Levanzin was also interested in spiritism and telepathy. He attended international conferences on Esperanto and ended his days in trouble for his heterodox medical practices (see Chapter 3).[53] As professional fasters challenged the longstanding religious tradition of abstinence, diet, and asceticism, spiritists also jeopardised traditional religious values. Equally, elixirs and mysterious liquors seemed to emerge from occult, often fictional experiences with dead spirits rather than from more established pharmacopoeias (see Chapter 5). It was again in that uncertain terrain that hunger artists gained credibility.

Psychology

During Tanner's fast in New York in 1880, the New York Neurological Society (NYNS) sent some of its members to oversee it and tried to discredit him for the lack of rigorous vigilance and potential fraud.[54] Additional doubts on the rigour of the supervision committees and suspicion of potential fraud did not help either. However, Tanner defended his assertion that the fast was both the result of willpower over the animal appetite and another proof of the virtues of eclectic medicine in order to prove the 'errors' of physiology. There was no clear closure of that public controversy, but it led to new explanations of the human resistance to inanition that again challenged the status of physiology as a unified, homogeneous discipline.[55]

In the context of the slippery physiological–psychological boundaries of the 1880s and 1890s – Wilhelm Wundt founded his Institute of Experimental Psychology in Leipzig in 1883 – the physiology of the senses (hunger in our case) and the study of the autonomy of the nervous system were linked to the causes of resistance to inanition. In fact, ten years earlier, with the introduction of the clinical illness anorexia nervosa by Ernst-Charles Lassègue and William Gull, the perception of abstinence from food shifted from the religious to the pathological.[56] The experimental culture that served physiology and nutrition (see Chapter 3) was also applicable to many of Wundt's experiments and to our doctor–hunger artist cases.[57] Richet, one of the leaders of the research programme for the scientific study of paranormal phenomena, attributed the capacity for prolonged fasting to states of hysteria and to the 'psychological' factors of resistance to inanition, and added new arguments in favour of a specific research project for humans compared with the traditional experiments with animals.[58] Suspicions of insanity shifted scientific explanations of the causes of resistance to inanition during fasts towards the realm of the human psyche. It was thought that hunger artists were able to substantially reduce their sensation of hunger in the first few days, but in the long term their resistance was supposedly based on a sort of hypnotic dream, autosuggestion, or willpower. Nevertheless, to be admitted as reliable objects of scientific experimentation, they had to be classified as 'normal'.[59]

In 1885, following instructions from the Italian consul in Cairo, Succi underwent psychological tests by Dr Bruno Battaglia, and the diagnosis was 'monomania or *paranoia ambiziosa*'.[60] In 1886, Monin and Maréchal admitted that Succi's and Merlatti's fasting power had to do with 'psychological' factors: 'We do not claim that Succi is a mad man, or our excellent Merlatti, but it is certain that they are *eccentric, nervous,* and *cerebral,* and, supported by an exaggerated and *hysterical* faith in their strength, eliminate the feeling of hunger and resist inanition by one of these phenomena of *auto-suggestion.*'[61] In 1888 in Florence, Luciani reported that Succi had been in the Lungara asylum twice,[62] and the diagnosis was '*frenosi sensoria*'.[63] Luciani also referred to Dr Lava from Turin, who defended Succi's 'paranoia', and reiterated Battaglia's remarks that hysterics, melancholiacs, and paranoiacs tended to practise voluntary fasts. Luciani also mentioned Lombroso's opinion – after Succi's fast in Paris in 1886 – who considered him a '*neuropatico*'.[64] Luciani's conclusions later appeared in his physiology textbooks and opened the door to psychological explanations of the functioning of the body. He stressed that: 'no matter how much I support the scientific or positive method, we must recognise that, on the one hand, it is impossible to explain specifically vital somatic phenomena only with the laws of physics and chemistry, and on the other hand, that psychic phenomena (of sensitivity and awareness), which are the culmination of life, avoid all mechanical explanation ... physiology cannot ignore the psychic phenomena'.[65]

Merlatti also suffered psychic disturbances during the thirteenth day of fasting, in Paris, as described in the report by Maréchal and Monin: 'This morning, the subject had a nervous attack which lasted approximately fifteen minutes, the symptoms of which were those of an attack of *hysteria*. This crisis was followed by quite an intense episode of dyspnoea: his eyes were bloodshot, and his face was puffy. An application of dynamometric plaques quickly conquered these events and showed that Merlatti, like all *neuropaths*, was susceptible to the highest point of *metallotherapy*. Merlatti claimed that he had suffered similar attacks in his younger years.'[66] After that nervous crisis, the medical committee advised him, unsuccessfully, like on other days of the fast, to stop his performance. In fact, Monin was convinced that the resistance to inanition was in the hand of characters and illnesses such as hysteria, hypochondria, mental alienation, and catalepsy. In Succi's case, and sharing the diagnosis of other contemporary doctors, Monin rejected the hypothesis of real insanity, but he considered him an eccentric, with an exaggerated faith in his own forces, which Monin called his '*aura sacra fames*'.[67] In 1888 in Florence, before his fast, Succi had a public reputation as insane or charlatan ('*matto*', '*simulatore*').[68]

French anatomist François-Achille Longet considered that hunger was a fully material phenomenon in which spiritual forces played no role at all.[69] Others, such as Marcel-Eugène Gley (1857–1930), an expert on the nervous system, professor at the Collegè de France, and member of the Academy of Medicine,[70] resisted any strict materialist explanation,[71] favouring the introduction of the role of the nervous system as the main explanation for Succi's resistance.[72] In 1890, the *Pall Mall Gazette* reported on Succi's own explanation of his capacity to resist inanition in the following terms: 'It is a power like *hypnotism*, like *galvanism*, a power, which I have gradually learned to create and to control. Solely by my will, I am able to abstain from all foods; it is power of will which keeps my brain clear and sustains me during these long fasts. *My will is stronger than my body; there you have the whole secret*.'[73]

Moreover, materialist positions were fiercely criticised by the well-known French parapsychologist Albert de Rochas (1837–1914), who was particularly interested in all sorts of paranormal phenomena (Figure 4.2). In his book *La suspension de la vie*, de Rochas devoted the first

Figure 4.2. Portrait of Albert de Rochas, 'Le colonel Rochas', photograph, 1895.

chapter to the prolonged fasts, again describing Tanner's pioneering fast and his imitators.[74] De Rochas presented cases of prolonged dreams, lethargy, and hibernation, alongside famous episodes of hunger artists (from Tanner to Succi), auto-experiments of fasting, and – as discussed in the next section – some cases of pathological fasting, associated with fasting women. De Rochas also appropriated the case of Succi and Merlatti for his own research agenda, which had led him to study the famous spiritist medium Eusapia Paladino.[75] By contrast, and close to materialism, science populariser Wilfrid de Fonvielle bitterly discredited Rochas and any spiritualist explanation of the extraordinary capacity of prolonged fasting in some individuals.[76]

In Nancy, Dr Hippolyte Bernheim (1840–1919) associated Succi's fasting capacity with his power of auto-suggestion.[77] He thought that the hunger artist's resistance was based on a sort of hypnotic dream, or willpower, and also opened the door to a range of psychological explanations. In fact, the École de Nancy defended the theory that hypnotism could provide a convincing explanation for the partial suspension of vital functions.[78] In 1886, the year of the Succi–Merlatti contest in Paris, Bernheim published his treatise *De la suggestion et de ses applications à la thérapeutique*,[79] which became in practice the manifesto of his research school; it was translated into German by Sigmund Freud in 1888, after his early studies on hypnosis. Opposed to the nosological, morbid approach of Jean-Martin Charcot's Salpêtrière School, Bernheim believed that hypnotism was a sort of natural dream, which made the brain more accessible to suggestion, and that it was therefore a useful therapeutic tool.[80] Bernheim described that, 'having suggested to two hysterical women, whom he had put to sleep, the absence of hunger and the order to not eat, [he] was able to subject them to a fast of fifteen full days, during which they drank, but did not ingest any solid food. This fast, which was well tolerated, could have been extended by another fifteen days, but one of the sick people had already lost 3.2 kg and the other 5.2 kg.'[81]

In a paper he devoted specifically to Succi, Bernheim pointed out how the hypnotic dream was another useful method to calm the feeling of hunger.[82] As the alienated, the hysterical, and the anorexic became insensitive to hunger and resisted much longer periods of inanition than healthy individuals, Succi would have practised a sort of autosuggestion.[83] Bernheim distinguished between *inanition* and *hunger* as two separate concepts. Inanition killed slowly and hunger very quickly. If hunger artists took their fast to the limits of human resistance, they would therefore die of hunger, not of inanition. In Bernheim's view, the pathological cases (insane people, anorexics, and hysterics) did not feel hunger. Opium, anaesthesia, chloroform, and the hypnotic dream were all strategies to kill the feeling of hunger and therefore to prolong life. Bernheim concluded that Succi 'is a believer, neutralising the sensation of hunger, fanaticised by his faith in the efficacy of his beverage, he does not die of hunger because he is not hungry'.[84]

In 1889, in his studies on human inanition and referring to Succi's and Merlatti's 1886 fasts in Paris, Charles Richet considered that they were perhaps alienated or melancholic.[85] Richet was referring to Succi's controversial stays in asylums at the beginning of his career as a hunger artist, but he put particular emphasis on the multiple cases of mental alienation, which led to increasing resistance to inanition. Richet presented records of fasting by Merlatti (50 days), Succi (30 days – before his 45-day fast in New York in 1890), and Tanner's legendary 40 days, with the 47 days of the '*amaurotique de Bérard*', the 63 days of the '*malade de Desbarreaux*', and the 76 days of the '*aliéné de Devilliers*'. He also added a long list of cases from the sixteenth century to the present time, when he worked at the Parisian Salpêtrière hospital, treating hysterical women with an impressive resistance to inanition.[86]

Richet again highlighted the role of the nervous system in the human capacity to resist inanition, and the different patterns observed in forced, voluntary, and ill fasting. Richet shared with Bernheim the idea that: 'A healthy man dies if he stops eating, after a certain number of days; a sick man remains without nourishment for several weeks with impunity.'[87] But they disagreed on the nature and therapeutic uses of hypnosis and on the definition of auto-suggestion. Unlike Richet, Bernheim defended the idea that auto-suggestion was a key explanation for hunger artists' resistance. In the case of Cetti's fasts, he admitted that the hunger artist was not a hysteric, but capable of auto-suggestion and of keeping all his physical force. Bernheim became firm in the belief that the visitors' stimulus and the hope of triumph were a sign of hunger artists' capacity to dominate their body. It was therefore the faster's will that provided his resistance to eating. However, Richet was more reluctant to accept that the human will could go as far as to achieve the suppression of the sensation or feeling of hunger.[88]

In New York in 1890, doubts about Succi's sanity continued. The *Pall Mall Gazette* reported that: 'Power of will, as he asserts, is really his whole secret ... Of his power of will we may read signs in his face.'[89] Although it had promoted and sponsored Succi's fast in the city, the *New York Herald* had to address the issue in its daily articles and reported the case in the following terms: 'He said that he had received certificates from doctors all over Europe testifying to his sanity, and though he had once been shut up in an asylum he never was in the slightest degree insane. The doctors examined him in order to make sure of his mental condition. They reported him perfectly sane, but a little light headed from confinement and bad air.'[90] Even in 1891, rumours about Succi's insanity still circulated in public with details of a new stay in an asylum, as had already been reported twice in his youth just after his African trips.[91] Again in the synchronic spirit of the land of the hunger artists, rumours about Succi's insanity soon crossed the Atlantic. In 1892 in Mexico, the press published an interview with Succi, sent by the Paris correspondent of *Il Corriere di Napoli.* The faster solemnly stated that: 'I think that I am cured and I recognise that I was crazy. There are probably two causes. For the six months I spent in London, I barely ate and only drank tea. When I got to Paris I was not well, I started to drink a bit of everything and I ended up becoming inebriated on more than one occasion … A friend of mine who is a hypnotist came with me from London and enjoyed hypnotising me … Perhaps in my tendency to go crazy there is the same power that I have to fast for long periods.'[92] It is particularly relevant to note that Succi placed his possible insanity and his willpower to withstand prolonged fasts on an equal footing, suggesting a sort of continuity between the two 'talents'.[93]

The study of the psychic factors of inanition and hunger fitted with the context of spiritists and the so-called psychical research battles around the nature of the emergent psychology and the unconscious.[94] In 1897, French doctor Henri Lassignardie published his *Essai sur l'état mental dans l'abstinence*, in which he studied the psychical problems in voluntary fasting in detail – the mystiques and the '*grandes jeûneurs*' (Succi, Merlatti), and involuntary fasting, in the case of '*des grands naufragues*'. Lassignardie considered that the psychical problems of inanition were very similar to those of intoxication.[95] This obviously rendered human inanition, if possible, in supervised laboratory conditions, a very useful potential case for the emerging science of experimental psychology. In 1905, a *Daily Express* journalist provided a good description of Riccardo Sacco's priorities of his practices and the psychological profile of the whole explanation: 'Riccardo Sacco informed me that he practises these public fastings, of which this is his twenty-fifth, not so much *to make money as to demonstrate the power the mind has over the body through auto-suggestion* … When Sacco condescends to eat he contents himself with *vegetables*. He passes his time during his confinement studying scientific and spiritual works.'[96]

As Francis Gano Benedict pointed out in 1907, fasting experiments with animals were very valuable for physiology, but much less valuable for 'human physiology'. In his view, only a complete study of hypnotic cases, professional fasters, and voluntary subjects could help to draw the line between 'normal' and 'pathological' fasting. In 1907, in his taxonomy of the different kinds of fasts, Benedict labelled as 'number two' that of the 'fasting of the insane'. He categorically asserted that: 'fasts of weeks, if, indeed, not months, have been observed in cases of insanity'.[97] He referred to much earlier research by Dr Émile Desportes, *Du refus de manger chez les aliénés*,[98] which appeared in French as early as 1864, and used contemporary words such as 'psychiatrist' and 'psychologist'. Benedict also added the case of 'fasting in hypnotic sleep', that is to say, processes of suspended animations (like the Hindu fakirs) or hypnotic suggestion for the study of nutrition.[99]

In 1923, when Sergius Morgulis published his book on fasting and undernutrition, the psychological turn to explain hunger artists' behaviour was widely accepted. Morgulis considered that it was very doubtful 'whether a healthy individual, with a vigorous demand for nourishment, would willingly endure the hunger sensation unless he was dominated by *some powerful emotion*'.[100] In his view, only professional fasters with their appearance of courage to impress other people, experts in autosuggestion, individuals with faith in therapeutic fasting, religious or fanatical martyrs, and crank reformers who saw 'in inanition a panacea for all ills of the flesh'[101] could withstand that kind of powerful feeling. All of this is convincing evidence that the success of hunger artists and their relevant place in laboratories and public shows did not escape the psychological turn of the last decades of the nineteenth century.

Gender

In 1902, Madrid newspapers reported on a fasting woman ('*ayunadora*'), a Miss Emmey, in the following terms: 'Miss Emmey. This famous faster, emulator of Merlatti, Succi, Tanner, Pappus, will perform her interesting experiments for the public of Madrid on Sunday 19th in the Frontón Central. At 10 pm on the night of the mentioned day, Miss Emmey will enter a glass urn, which will be sealed, and she will stay inside for seven days without having any food. During this time, she can be visited by the audience at any time of the day or night.'[102] Not much is known about the itinerant adventures of Miss Emmey. If details of male hunger artists are hard to find in historical research, the case of women is an even more uncertain terrain. Moreover, beyond the weaknesses of that prosopography, in our period of study, contemporary actors approached hunger and inanition with a strong gender bias, often associated with specific psychological readings. In spite of the scarcity of sources, short newspaper articles are very informative on the cases of women fasting for varied reasons, from religious asceticism to neurasthenia to anorexia nervosa, or just as professional fasters themselves. These cases were popular and numbered, and their classification often hard to establish.

In our period of study, fasting women were often perceived as patients, whereas men became heroes for controlling their feelings of hunger and their athletic strength. While many considered male professional fasters as unique individuals with extraordinary physical skills to withstand long periods of inanition, women's fasting and abstinence were very often related to the pathological, to be read as a peculiar female obsession or, in other cases, associated with hoax and fraud.[103] Many male doctors perceived anorexia nervosa as a female form of hysteria,[104] whereas the male ability to achieve long-term starving was supposedly a demonstration of extraordinary willpower.[105] German feminist writer Hedwig Dohm reflected, for example, on female oppression in sanatoriums and their forms of resistance such as self-starvation.[106] In the United States, in the period of social activism and political reform – which roughly coincides with the hunger artists' times of splendour – fasting was mainly associated with values such as health, vitality, longevity, and the reinvigoration of the body (and society), always framed in a masculine world view.[107] The medical description of anorexia nervosa in 1873 by Lasègue and Gull became a key landmark for the pathologising of female fasting.[108]

Well-studied cases of fasting women such as Sarah Jacob, Louise Lateau, and Mollie Fancher provide useful data to refine a gender analysis of fasting, alongside other, later cases of professional fasters such as Auguste Victoria Schenk. The latter, escaping from the stigma of the pathological, played a more masculine role in terms of the commodification of the performance and its subjection to scientific study. There is also evidence of other women's fasts, which took place in the golden age of hunger artists, such as Kate Smusley (1884), Josephine Marie Berard (1888), Helen Coppague (1897), Mary Davenport (1904), and Claire de Serval (1910).[109] As happened with William Hammond's opposition to Tanner's fast, controversy around fasting women became a powerful tool for doctors to draw their own professional boundaries, areas of expertise, and research priorities not only in the academic arena, but also in the public sphere and particularly in the daily press. Further details on some of the most relevant women fasters will help us to depict more precisely the gendered dimension of the metier.

Sarah Jacob (1857–1869), known as the 'Welsh fasting girl', became a very popular case as a result of her claim to have withstood starvation for long periods. In 1869, a medical committee supervised her to avoid any ingestion of food, and after two weeks, she presented clear signs of starvation. Her parents' refusal to stop the fast led to her death. Later, it was proved that she had secretly consumed food, so the whole story became fraudulent, and the parents were prosecuted.[110] In 1871, Dr Robert Fowler, a member of the Hunterian Society and its president in the 1880s, used the case to define boundaries between supposed charlatanry and the academic authority of an empirical medicine.[111] He thoroughly documented his study with doctors' controversial views, reports of medical committees, trials in the courtroom, and details of Sarah's death. Fowler concluded the case could probably be associated with 'an anomalous form of Nervous Disorder', similar to catalepsy and close to ecstasy, or perhaps it was a case of 'simulative hysteria'. He also reported that during the fast, Sarah was seriously emaciated, and became almost skeletal. Fowler blamed the parents, since they supposedly 'were cognizant their daughter did not live entirely without food'.[112]

Other fasting cases in women also brought public discussion towards the pathological terrain. This was, for example, the case of Louise Lateau (1850–1883), whose stigmata and mystical trances, accompanied by long periods of fasting, drew the attention of physicians, members of the clergy, and the press from the 1860s to the early 1880s. Again, the case was appropriated differently by numerous contemporary witnesses. Clergy read Lateau's case in a religious framework. Catholic doctors tried to find a balance between religion and experimental physiology. A physician from Louvain, Ferdinand Lefebvre, concluded that Lateau's stigmata and her mystical trances were miraculous in nature. Other scientists aimed to find more 'objective', 'rational' experiments for the analysis of the case and the control of fraud. Lateau's case was also examined in the Belgian Royal Academy of Medicine. For many others, Lateau was considered another case of hysteria, in tune with the gendered psychological turn of the last decades of the century.[113]

As a reaction to Lateau's tremendous success in popular literature,[114] the prestigious German physiologist Rudolf Virchow was asked to study the case 'scientifically' and to make his conclusions public. An apology about Lateau's case by Catholic theologian Augustus Rohling,[115] which sold around 50,000 copies and received support from other doctors, put more pressure on Virchow. After some hesitation, Virchow invited Lateau to his laboratory at a Berlin hospital, but she refused. Virchow's materialism and anti-clericalism explain his position demanding further everyday surveillance of Lateau, something that could easily be done in a laboratory or in a hospital rather than at her home, again to avoid reasonable suspicion of fraud. In Virchow's view, such a long period of fasting fully contradicted the physiological knowledge of the time.[116] The Lateau case led to longstanding controversies among the medical profession. Some members of the Belgian Royal Academy of Medicine supported Lateau's extra-sensory powers and fasting capacity, but it provided new intellectual tools for popularisers such as Fonvielle in his crusade against pseudo-science and charlatanism, and for his idealisation of rational, scientific minds such as Virchow's. At the end, though, the whole issue turned again to the pathological with the theory about the capacity of mentally ill patients, particularly women, to have an extraordinary resistance to inanition.[117]

The tension between gender, insanity, and hunger was also intrinsically linked to Henry Tanner's own biography and experience with fasting. After his first attempt at fasting in 1877 in Minneapolis, Tanner soon became involved in the public controversy of the case of someone known

Figure 4.3. Mollie Fancher, known as the Brooklyn Enigma, in a trance in 1887.

as the Brooklyn fasting girl, Mollie Fancher (Figure 4.3). In 1865, after suffering an accident, she could not see, touch, taste, or smell, but began to predict future events, use clairvoyant powers, and claim periods of abstinence from food of fourteen days. Her home became a very popular attraction, but again she was discredited as a case of fraud and hysteria. Tanner gained public relevance through his defence of Fancher's case, in a bitter controversy with Dr Hammond, the influential member of the NYNS, who opposed eclectic medical practices and dismissed Fancher's scientific credibility.[118] Members of the NYNS supervised Fancher's state and discredited her in the press. In 1879, Hammond himself published his book *Fasting Girls: Their Physiology and Pathology*,[119] in which he described Fancher's case – again – as fraud and hysteria. Hammond considered that trances (as in Lateau's case), hibernation, hypnotism, and self-induced catalepsy were often related to the ingestion of drugs such as chloral, morphine, or ether.[120] However, Tanner disagreed, challenging Hammond and the NYNS to empirically 'test the powers of human endurance under prolonged fasting, or witness the physiological, pathological, or psychological phenomena incident to such a fast'.[121]

Tanner's disagreements with Hammond on what was known as the 'Brooklyn enigma' of Mollie Fancher made him better known in public and helped the further organisation of his famous fast in New York in 1880. Tanner justified his interest in Fancher's case as part of his opposition to a one-sided medical legislation, and in coherence with his belief in primitive Christianity. Tanner's belief in spiritism and in the hegemony of willpower over the animal appetites facilitated the appropriation of Fancher's case into the spiritual, psychological realm.[122] The *New York Herald* published several articles in favour of a 'scientific' test to assess Fancher's clairvoyance and her actual resistance to starvation.[123] Tanner wrote to Hammond and to the president of the NYNS, Professor Joseph R. Buchanan, but he received no answer.[124] In spite of the tensions, Tanner contributed to the legitimation of women's fasting as part of his aim of promoting the therapeutic values of abstinence and diet. In fact, in the late 1890s, Linda B. Hazzard followed Tanner's advice for her own fast, as part of her campaign in favour of a natural therapy without drugs. Hazzard was also trained under the guidance of Dr Edward H. Dewey, another of the great advocates of the virtues of diet and abstinence.[125]

Controversies around cases such as those of Jacob, Lateau, and Fancher easily blurred the boundaries between fasting, religious mysticism, science, gender, and spectacle.[126] They also contributed towards spreading gender-biased medical explanations of the fasting capacity of some women, and suspicions of fraud and charlatanism. Fraud plus hysteria became two key concepts to accuse many of these women of being psychologically ill and to raise suspicion. Nevertheless, in other cases, some professional female fasters played their cards imitating the role and strategies of their male counterparts. They stressed their own 'male' virtues of inner resistance to inanition such as heroic performances, strong physical resistance, and the undoubted scientific interest of their cases. In 1892, the female hunger artist Alma Nelson fasted in Paris. Born in New York to a poor family, she was raised by an 'Indian' nanny who knew the therapeutic secret of exotic plants and used them to cure the girl's illnesses. Nelson spent her youth in Paris singing in coffee shops and restaurants and self-experimenting in periods without food. Her performance in 1892 followed the standard pattern of a professional faster selling her own commodity in a secular marketplace.[127]

In other cases, like male hunger artists, women were subjected to medical experiments during their fasts and acted as professional fasters in terms of the public exhibition of their emaciated bodies in the marketplace. This was the case of actress Auguste Victoria Schenk, who subjected herself to experiments in physiology laboratories in Hamburg,

Vienna, and Berlin as a professional fasting woman.[128] In 1905, reproducing some aspects of the 1886 Succi–Merlatti contest in Paris, Schenk challenged male hunger artist Riccardo Sacco, who had fasted for twenty-one days in a coffee house at the Wiener Prater. Schenk's attempt to fast for twenty-three days was praised by the *Illustrirtes Wiener Extrablatt*, which celebrated her as the 'greatest phenomenon of the twentieth century', but at the same time the newspaper ridiculed her desire for emancipation with comments such as 'even the weaker sex can have a strong stomach'.[129] Schenk's fast in Hamburg in 1905 provided data for further experiments in a medical clinic in Berlin, with the publication of three papers in the *Zeitschrift für experimentelle Pathologie und Therapie*.[130]

In 1898, an anonymous woman hunger artist also appeared in the West Court of the Royal Aquarium London to begin a thirty-day fast and receive visitors daily. She shared the stage with the Australian Giants, the Harem and Zeo's Maze, Fern, Fay, Sybil and the Mermaid, Professor W. J. Cook, phrenologist, and a 'Wonderful Series of Animated Photographs', projected in a 'Lumière's Cinemaphotograph', as a sign of the spread of cinema across the Channel just three years after its invention by the Lumière brothers in 1895.[131] After the failure of this fast, Mme Auguste Christensen took the risk again in 1901 in the same venue. She had already fasted for fifteen days in Schleswig (drinking only tea and coffee), fifteen days in Hamburg (drinking only water), and twenty days in Copenhagen (again with tea and coffee alone). Now the fast at the Royal Aquarium was planned for thirty days and had to be supervised by three doctors, four nurses, and four female overseers.[132] Another professional female faster, Flora Tosca – a quite unknown, almost anonymous figure – received this epithet as a tribute to Puccini's opera, composed in 1899. She became the object of detailed scientific research, which was published in 1905–1906, in different medical journals.[133] Also in the laboratory context, in 1907, Benedict published a thoroughly documented paper on the composition of the urine of a starving woman, which he considered a new contribution to the observations made on a female patient of the Connecticut Hospital for the Insane, together with seven other 'insane' women.[134] Several physiologists also ran studies on fasting women.[135]

In 1909, a self-experiment by the medical student Claire de Serval was also of particular interest. Recovering the old tradition of the therapeutic properties of abstinence from food, Serval considered that many illnesses were due to overeating and could be cured through short periods of fasting, at least two days per week. She voluntarily enclosed herself in a glass cage with small ventilation windows that prevented the possibility of any food being introduced from outside. The experiments took place in the Charity Hospital of Berlin, with a respiration calorimeter. All physiological functions remained stable. After twenty-four days of fasting, Serval had lost 7 kg (291 g per day), small when compared to Succi's case. Serval achieved a forty-day fast, maintaining good physical condition. Serval left a very rich set of personal letters in which she fully described her main self-experiments, often with the collaboration of some medical overseers in Berlin. Albert de Rochas's report in *La suspension de la vie* (1913) was particularly enlightening:

> In Berlin, Mlle de Serval, who has done serious medical studies and who considers most of our illnesses to be due to overeating and irregular eating, became cured of several illnesses through short, strict fasts of two to six days, during which she drank only pure water; she believes that everyone (young and old alike, the poorly and the healthy) should fast for two days a week. To enable the doctors to study the physiological effects of hunger, she voluntarily enclosed herself several times in a glass cage measuring 3 metres in length by 2.5 in width and 2 metres in height, hermetically closed on the four corners … Despite the rather poor hygienic conditions of a narrow cage to stop all exercise, all the physiological functions remained perfectly normal … Despite the pale tone of her skin, the doctors observed, in the case of Mlle de Serval, a remarkably constant level of haemoglobin in her blood … The last and longest fast by this lady was forty days, during which she consumed no food and only consumed a small amount of pure water (instead of the mineral water sometimes used). The letters written during her voluntary captivity bear witness to her perfect lucidity and the activity of a highly cultivated mind.[136]

There is no doubt that contemporary witnesses had a gender-biased approach towards women's fasts, often associating these cases with examples of psychological illness and potential fraud. Nevertheless, cases of professional women fasters, who followed the pattern and procedures of their male colleagues, as well as the scientific study of women during starvation, are clear signs of the relevant role that female fasters played in the whole endeavour, and how they challenge a too simplistic view of a supposedly well-defined boundary of the art of fasting from a gender perspective.

***

Beyond a too simplistic, polarised approach to the different physiological schools, there was a huge range of grey zones, suitable for hunger artists. In Barcelona, during Succi's fast in 1888, some members of the supervision committee strongly opposed any spiritualist explanation for his resistance to inanition. Although enthusiastically encouraging experiments with hunger artists as a crucial practice in modern physiology and nutrition, Dr Avelino Martín argued, for example, that Succi was led to that series of experiments by a totally incorrect doctrine, mixing rationalism and spiritualism.[137] Martín perceived hunger artists as an excellent opportunity to overcome limitations of animal experimentation, but dismissed any non-materialist explanation.[138] However, in the same local context, other doctors had a different perspective. At the Academia Médico-Farmacéutica de Barcelona, Dr Javier de Benavent was again in favour of the vital force and auto-suggestion.[139] Martín finally admitted that a plausible hypothesis to explain Succi's resistance to inanition could be a certain paresis of the nervous system stimulated by auto-suggestion.

The case of Barcelona can also be extrapolated more generally. In the last decades of the nineteenth century, physiology did not provide a holistic, comprehensive explanation of the functioning of the human body, nor could it avoid trends of medical specialisation. Limitations of experimental physiology when attempting to find convincing explanations for the causes of human resistance to inanition in prolonged fasts paved the way for the leaking of heterodox 'spirits', and for the emergence of the psychological turn. Vitalistic trends opposed materialism to different degrees throughout the century, and they never disappeared. Criticism of materialism also became useful to explain hunger artists' impressive resistance and opened the door to more 'spiritualist' explanations. In a context of medical heterodoxy, even doctrines such as spiritism found their place in the land of the hunger artists, especially in the case of Succi and his construction of a spiritist universe, which supposedly had its roots in his African trips. Were hunger artists heroic individuals, monsters with exceptional qualities who explored the resistance of the human body, or perhaps just insane, ill individuals, with the additional suspicion of fraud? Answers to these questions led to intersections between the practice of professional fasters and the faith in the heterodox doctrine of spiritism.

In their performances, hunger artists became mediums. Unlike those (women) who supposedly brought the spirits of death into contact with the living, our professional fasters 'mediated' between medical and popular knowledge on elusive concepts such as hunger, appetite, inanition, and starvation. They also provided fruitful terrain for the growth of psychological explanations behind the human capacity of resistance to inanition, from willpower to insanity and hypnotism. From Kardec's spirits and Succi's spirit of a lion, to the psychological explanation of the Nancy School of hypnosis and Charles Richet's considerations, the study of hunger artists contributed to moving the boundaries of experimental physiology and to opening a new branch of psychological explanations. Not by chance, the psychological turn of the last decades of the nineteenth century coincided with the golden age of our professional fasters.

From a gender perspective, the supposedly male, athletic heroism of professional fasters, who were now already disentangled from religious, mystical considerations and who were able to reach the limit of human resistance with their spectacular prolonged fasts, had an obvious gender bias. Cases such as those of Sarah Jacob, Louise Lateau, and Mollie Fancher, often on the fringes of religious beliefs and diagnosed as pathological cases of anorexia nervosa or hysteria, and labelled as 'fasting girls', were more often tinged with suspicion of fraud than their male colleagues. Nevertheless, other women acted as professional fasters, imitating the masculine pattern of that commodity, undergoing rituals of medical and civic supervision. In times of heterodoxy, fascination with the occult and its scientific study, spiritists, mediums, hysterical patients, and professional fasting women also crowded the land of the hunger artists and the physiology of hunger and inanition, in a permanent crisis of unity and disintegration.

5 Elixirs

Apart from fasting, Henry S. Tanner's eclectic therapeutics also included plant remedies, a moderate ingestion of food, the use of hydropathy and other 'natural' treatments, and a special emphasis on individual lifestyle.[1] Tanner was convinced that air and water (a kind of late neo-Hippoçratism), electricity in the atmosphere, and regular abstinence from food worked better than allopathic drugs as therapeutic strategies,[2] and refrained from drinking coffee, tea, and liquors and from consuming tobacco.[3] In the context of American eclectic medicine, in which Tanner developed his fasting skills, heterodox medical practices such as vegetarianism, naturism, homeopathy,[4] and the use of elixirs and mineral waters raised public interest.[5]

In fact, Tanner's therapeutic strategy fits well in a more general picture of his times. Despite the increasing medical professionalisation throughout the nineteenth century, charlatans and peddlers had a good number of controversial drugs in their hands. Since legislation generally failed to give a medical monopoly to regular practitioners, competition for patients remained severe in a fierce struggle for authority and social acceptance.[6] These drugs – often known as 'patent medicines' – did not usually treat specific illnesses, but instead acted as supposedly general 'miracle elixirs', often containing opium derivatives. In addition, beyond an 'orthodox' medical culture that was often offset by very critical, aggressive campaigns, the drugs were also part of the contemporary culture of entertainment, hype, and fraud, and in freak shows, flea circuses, or magic tricks, which, as we already know, also included hunger artists.[7] Elixirs provided additional arguments for speculations on trickery, so hunger artists' liquors and the mineral waters they drank were relevant agents that deserve further investigation.[8]

Controversial drugs were usually patented under a commercial name, but without providing adequate details of the formula. To add extra appeal to these commodities, their exotic origin was presented in public as the result of colonial travels to remote places on other continents. Advertisements also had doctors' endorsements of the supposedly universal curative properties of the liquors.[9] In other cases, local herbs provided genuine, original formulas with supposedly specific curative properties. Among the 'miraculous elixirs' one could find for example cocaine, Indian oils – supposedly based on Native American traditions, which were very popular in Tanner's eclectic medicine – cannabis, arsenic, the famous Dr Pepper's tonic,[10] and Dr Williams' Pink Pills, the latter made of iron oxide and magnesium sulphate to treat anaemia.[11] That elixir culture also ran in parallel with the expansion of the chemical industry, which was able to produce synthetic drugs from coal tar, but at the same time faced uncertainties about the composition of complex mixtures of natural products of animal and plant origin.[12]

In practice, in the market of medicines, which ranged from drug peddlers and charlatans to apothecaries and university doctors, all actors were seeking profit, social recognition, and scientific authority. In addition to this, the culture of drugs and elixirs endorsed other controversial medical practices such as homeopathy. In some places, its therapeutic treatments, also opposed to allopathic medicine, became ideal weapons in the battlefield for the use of hunger artists by eclectic practitioners. Moreover, like hunger artists and their elixirs, mineral waters also became lucrative commodities in the context of the spa culture and the hydropathic cures of the time. Water was sold and advertised seeking the legitimation of a medical authority, and often with the collaboration of professional fasters themselves, who acted as distinguished patients and consumers and as marketing agents.

Once hunger artists stopped eating solid food, but drank distilled water or any other mineral water, and ingested drops of elixirs, drugs or 'mysterious liquors', further controversy and commotion spread in academic circles and throughout public opinion. Paradoxically, though, disagreements on the chemical composition of elixirs and mineral waters and on their therapeutic properties during a prolonged fast made scientific consensus harder, but again boosted hunger artists' influence in public debates and increased their aura as celebrities. Potential consumers of all these liquids wanted to know more about their reliability and therapeutic properties. So, now, in addition to the authorised medical voices, there were many moving, personal experiences of professional fasters, the 'heroes' of human resistance to inanition, whose journeys and performances were regularly accompanied by precious bottles of all sorts of liquids to be sold. The growing influence of eclectic therapeutic treatments, the commodification of hunger artists' elixirs, and the challenges posed by their chemical composition were signs of the fluid, unstable terrain of the land of the hunger artists, a place in which uncertainty and charlatan-like fraud went hand in hand with the liquids, and enlarged the battlefield for scientific authority.

Homeopathy

Just as hunger artists were often dismissed by contemporary 'orthodox' science, many medical doctors, often coming from the most materialistic schools in experimental physiology, fiercely attacked homeopathy. The tensions between allopathic and homeopathic practitioners reached the public sphere, even the domestic circles in which women played a very significant role. In that context, a certain 'unorthodox professionalisation'[13] gained ground among homeopaths and created a sort of extended battlefield for legitimation from university chairs, medical societies, public lectures and popular pamphlets, with notable analogies with hunger artists' fringe culture, from laboratories to showcases and the daily press. In addition, that battle to recognise homeopathy and its links to public fasting often had a very genuine local character.

In 1888, for example, a considerable number of the Barcelona homeopaths belonged to the medical board of Succi's fast and perceived the case as an excellent opportunity to reinforce their medical doctrine before the hegemony of allopathic treatments.[14] Dr Javier de Benavent was president of the board, and member of the Ateneu Barcelonès, an institution to which he introduced Succi before the latter began his public fast. His deep homeopathic convictions acted as a useful weapon in his passionate defence of a vitalist physiology. In fact, the homeopathic corpus intersected with longstanding vitalist trends, which, as we already know, seldom disappeared from the medical culture when approaching hunger artists.[15] This homeopathic 'vitalism' introduced abstract, elusive concepts such as 'spiritual', 'psychic', and 'vital principle', which were familiar in narratives about hunger artists' performances and reinforced their psychological turn (see Chapter 4). Benavent, who also sold his own homeopathic drugs in pharmacies around the city and advertised them in the general press,[16] concluded that Succi's case was an excellent example to help discard the strict materiality of the human body.[17] He even suggested that a sort of inner spiritual force controlled the dynamism of life in the same way that homeopathic drugs stimulated the body's resistance to illnesses.[18]

In Benavent's words:

> during health, a spiritual force governs the organism, maintaining harmony in it, for without this vital or spiritual force the organism cannot live, for in illness the vital force is only disharmonised primitively in a morbid way, expressing its suffering by anomalies in the way the organism works and feels, for in the course of any illness, whether physical or moral, it is useless to know how the vital force produces the symptoms ... Doctors, and even non-doctors with materialistic or positivistic ideas, do not understand this, just as they do not understand the way *our almost always infinitesimal medicines work*, accustomed as they are to seeing only substantial material forces in everything.[19]

From the 1870s onwards, including the golden period of the hunger artists, homeopathy was not a monolithic system.[20] In his *Organon of the Healing Art* (1810), Samuel Hahnemann's main principles had paved the way for a huge range of medical practices. Homeopaths could, for example, assimilate some aspects of the new laboratory medicine, but differ in their specific medical views in different local contexts. Homeopathy could also provide new explanations for the exceptional cases of individuals resisting prolonged fasting and taking specific drugs, such as Succi's liquor. The individualistic understanding of the illness could be associated with hunger artists' singularity as patients, with their rare physiological resistance to inanition. Moreover, the homeopathic assumption of the unity of the organism and the trust in the self-healing powers of nature were not far from the extraordinary conditions and willpower associated with professional fasters.

In 1885, Wilhelm Ameke (1847–1886), one of the main followers of Hahnemann's doctrines,[21] reminded his colleagues of the old tradition that, in the hands of a skilled physician, potential poisons could become beneficial drugs and easily decomposable substances could be converted into powerful remedies.[22] In the 1880s, this idea could affirm doctors' skills to prepare new drugs, but also hunger artists' talent to deal with mysterious liquids. The homeopathic small-dose therapy was in tune with hunger artists' ingestion of small doses of liquors and mineral waters as the only liquid they took during the fast and reinforced the homeopathic idea of simplicity in prescription. The need to test the effect of drugs tacitly sought professional control of the therapy, which was also not much different from the medical boards and their strict supervision of public fasts. Finally, homeopathic treatments provided a curative power to the diseased organism, just as willpower provided a strong resistance to inanition in hunger artists.

In 1888, in a paper in the journal *El Consultor Homeopático*,[23] one of the most prominent journals of the homeopathic medical community in Spain at the time, Dr Salvador Badía, another member of the Barcelona medical board of Succi's fast, pointed out the unique scientific interest of Succi, which transcended the commercial nature of the performance and its entertaining aspects. In his own words: 'for us he is an extraordinary man who is solving one of the most interesting problems of therapeutics; that is to be able to sustain the individual for a certain time in order to cure many diseases of the stomach, liver, and nutritional centres that need abstinence in order to achieve a cure … what makes Succi's experiment remarkable is that it is a physiological phenomenon carried out without entering the field of pathology'.[24] Badía attempted to link Succi's fast with therapeutics of the homeopathic doctrine, and in doing so he reinforced his scientific authority as a practitioner of that 'heterodox' branch of medicine. His main theory was aligned with the contemporary tradition of therapeutic fasting already mentioned, which considered excessive food ingestion as the cause of numerous illnesses.[25]

At least in the Spanish medical culture of the time, the homeopathic approach to hunger artists grew on the fringes of spiritism, hydropathy, and other eclectic practices. Dr Anastasio García López, a distinguished homeopath and a member of the medical board of Succi's fast in Barcelona, had also attended the International Spiritist Conference that took place in the city that same year (see Chapter 4). García López was spiritist and anti-materialistic, a member of the Sociedad Hahnemanniana Matritense, and president of the Sociedad Española de Hidrología Médica.[26] His *Lecciones de medicina homeopática* (1872) was a key textbook in the field in Spain.[27] García López's eclecticism – à la Tanner – which moved from homeopathy towards hydropathy and vegetarianism, provided an ideal position for the appropriation of public fasts. Along similar lines, in late nineteenth-century Germany, 'fringe' medicine also included homeopathy, mesmerism, hydropathy, vegetarianism, and quackery.[28] Many doctors often accepted homeopathy as a potential treatment for numerous illnesses, and progressively joined 'alternative' routes of the medical culture. All these eclectic practices were often associated with issues related to fasting and inanition.[29]

Other events reinforced the link between homeopathy and the hunger artist culture. At the beginning of Succi's fast at the Royal Aquarium London in December 1890, the press assumed that his mysterious elixir had something to do with homeopathic medicine. As described by the *Globe*: 'On the present occasion [Succi] reserves to himself the right to take *occasional homeopathic doses of a mixture which he is ready to submit to any medical man for inspection or analysis, and in which he declares there is not the least nourishment*.'[30] In fact, in that freak show context, Succi's liquor was associated with a homeopathic drug. In addition to that, articles on the physiology of nutrition (and inanition) and the role of hunger artists in that medical context frequently appeared in homeopathic journals. In 1896, for example, *La Clinique. Organe de l'homeopathie complexe* included a paper on fasting and stomach illnesses, and mentioned the exceptional constitutions of Tanner, Succi, and Merlatti.[31]

In other cases, doctors reluctant to adopt homeopathy, but interested in issues of fasting and inanition, progressively changed their mind. This was, for instance, the case with the Argentinian doctor Luis C. Maglioni, who, in his youth in 1878, had submitted a PhD at the Faculty of Medicine in Buenos Aires that clearly opposed homeopathy. In his dissertation he pointed out that his main intention was to: 'address [himself], in comprehensible language, to the common people, who are the only ones in whom the homoeopathic system takes root, because the sap supplied by ignorance is the only one suitable for its nourishment'.[32] Some years later, though, in London in June 1915, following Tanner's tradition, Maglioni became a doctor faster. He spent thirty-seven days without food, but drinking water, taking homeopathic drugs, doing moderate exercise, and sunbathing. After the fasting experiences, in 1920, he even promised to add homeopathic drugs to his arsenal of therapeutics.[33]

Liquors

Giovanni Succi's liquor – a beverage that he ingested at the beginning of his fasts and that had some uncertain, exotic origin – caused continuous controversies, disagreements, and even contradictory interpretations of its nature and functions. To what extent was Succi's liquor just another fraudulent drug that itinerant charlatans used to sell to their mesmerised audiences? Or perhaps it was a tremendous challenge for the scientific authority of medical doctors and chemists who tried to link its puzzling composition to hunger artists' capacity to remain healthy during prolonged periods of fasting? These are undoubtedly relevant questions that reached the popular realm of Succi's contemporary witnesses. Giovanni Chiverny, the president of the medical board of Succi's fast in Milan in 1886, referred for example to a popular sonnet that spread across Italy and praised the heroic nature of the faster: 'Instead of eating, he drank liquor, which is not available in the tavern. Iron-strong faster! I salute you.'[34] From its exotic, mysterious African roots, to controversies around its chemical composition and its ambiguous nature, between food and drug, Succi's liquor – but also other liquids ingested by hunger artists during their fasts – provoked medical controversies about its actual function in the physiology of nutrition during the fast.

There were huge disagreements on the composition of the liquid.[35] Laudanum, chlorodyne and osmazone are probably the three key words to bear in mind. Laudanum was the medical form of morphine (an alkaloid of opium) that became very popular as a huge 'safe ward against pain, poverty and boredom', which was mainly used to fight pain and coughs. It was sold by charlatans and prescribed as a medical treatment.[36] Chlorodyne was a mixture of laudanum, cannabis, and chloroform, which was used to fight cholera, diarrhoea, neuralgia, and migraines. Osmazone, however, was an aqueous meat extract, which had nutritious properties as an additive to a specific diet. So, in that context, boundaries between food and drug, depending on the formula, marketing campaigns, and curative rhetoric, became blurred and complicated the consensus about the role of these liquors during prolonged fasting.

Succi's liquor's fame was associated with his travelling adventures and even reported by other explorers such as Georges Revoil, who, in 1886, led an expedition to the Great Lakes, Victoria, Tanganyika, and Malawi.[37] Being well aware of Succi's trips in Africa, Revoil considered that the former's liquor could be the same substance ingested by the long processions of people, a sort of 'raw opium loaf called kassoumba ... Thanks to this opium, stomachs do not call out famine at critical times.'[38] Revoil compared Succi to the ivory trafficking processions in which the bearers suffered from famine during the long journeys across unknown territories, and they supposedly numbed the feeling of hunger with kassoumba. Some other sources describe Succi's liquor as composed of cocaine, arsenic, extract of cucumber or *carica papala*, from which one could obtain osmazone, which would mean food – protein – and therefore fraud during the fast.[39] As described in 1886 in *La petite presse*: 'that is why Succi would not allow the analysis, because the chemists would discover osmazone, a meat or *carica papala* extract'.[40] Dr Luigi Buffalini, who had accompanied Succi in his fasting performance in Milan in 1886 as a member of the medical board, considered that Succi's elixir could be a fruit from equatorial Africa,[41] probably the 'kola nut': 'a kind of astringent chestnut, much appreciated by the people of Central Africa for its restorative properties, and enabling travellers to withstand food deprivation and long walks under a burning sun'.[42] Buffalini believed that certain drugs, such as opium (morphine), used in medicine, could suspend the feeling of hunger, but did not have any nutritious nature. In his view, coca, alcohol and chloroform had similar properties. Buffalini regretted the lack of knowledge of African flora but, precisely that fact convinced him to accept the 'miraculous' medical properties that Succi claimed his elixir had, and the doctor speculated about the possibility that the liquor helped Succi to retain very good vital signs during his prolonged fast (Figure 5.1).[43]

Fame and controversy around Succi's liquor also spread in the press. In 1886, in a tabloid style, the *Wiener Caricaturer* described, for example, the amazing properties of the liquor to provide strength to not eat delicious food and at the same time not risk the faster's life: 'The Succi Liquor/Succi, the great scholar/Invented a famous liquor/And whoever drinks just a little of it/He's definitely not hungry anymore/You look for schnitzel and roast beef/For goulash and *Beuschel* no more/And neither for pasta and strudel/With this brilliant liquor.'[44] The same year, the *Chemist and Druggist* acknowledged Succi's success at drawing doctors' attention and speculated on the presence in his liquor of arsenic, a

Figure 5.1. Detail of Albert Robida's caricature of Succi's liquor.

substance that had the property of 'producing in the stomach a warmth, which temporarily allays the pangs resulting from the deprivation of nourishment'.[45] In 1888, the liquor was sold in a shop in the Piazza della Signoria in Florence, and advertised in *Il Corriere Spiritico* – Succi's spiritist journal – as a drug with a potentially huge range of benefits.[46] Like other patent medicines, it was advertised in the following terms as a panacea with an impressive, exaggerated therapeutic capacity:

Succi Liquor. Florence. Piazza della Signoria, 7 (corner of Via dei Magazzini). Succi Liquor is medicinal and is composed of plant substances from Africa and was used by the inventor on his travels to Equatorial Africa. It is a marvellous medicine with a tonic action; it is in all respects prophylactic and therefore has an immediate effect in the following diseases: cholera, dysentery, diarrhoea, colic, persistent migraines, vertigo, hysteria, chorea, neuralgia, tetanus, eclampsia (pains suffered by pregnant women), rheumatism, dyspepsia, acidity, pain, intestinal atony, epilepsy, asthma, delirium of drinkers, heart defect, melancholy, etc. It is an excellent prophylactic against malaria fevers and is effective in soothing and stimulating. It also works for people suffering from insomnia and seasickness. It is an excellent tonic, using three or four drops with wine or vermouth. This liquor is one of the most excellent anti-spasmodic substances, so much so that in the first days of fasting Succi uses it to cushion the painful sensation of hunger, as was clearly demonstrated in the various experiments he carried out, among which we will name those in Cairo, Forlì, Milan, Paris and lastly Florence in March 1888, carried out under the supervision and studies of the Accademia Medico Fisica.[47]

In addition, the marketing strategy went hand in hand with the patenting of the liquor and its bottle, also in 1888, as it appeared in full detail in the *Gazzetta Ufficiale del Regno d'Italia*:

> I. Octagonal shaped bowl with the inscription *Liquore Succi* on the glass.
> II. Stamp and red sealing wax with the inscription: *Succi-Firenze.*
> III. Wet red stamp, with the inscription: *Liquore Medicinali Succi Firenze, Piazza della Signoria*, 7.
> IV. Wet stamp with facsimile of the signature: *G. Succi.*
> V. White label with the following text: *Liquore del Digiunatore, Esploratore Giovanni Succi*, etc.
> VI. Packing sheet in the background of which is the portrait of Succi between two lions and the inscription: Liquore etc.
> VII. Strip of green paper with the same emblem and with the inscription G. Succi etc.
>
> Said trademark or distinguishing feature will be used by the applicant to distinguish the special liqueur manufactured by him and entitled to his own name.[48]

Not much is known about the actual circulation of Succi's liquor in terms of a commodity or about its trade figures. But his ambitious marketing campaign, including its patent, and the experts' continuous disagreements about its composition put it on an equal footing to patent medicines and charlatanesque strategies for selling elixirs with supposed healing powers. In fact, the examples of other hunger artists showed a similar profile. In 1886, Alexander Jacques fasted at the Hotel Dieudonné, Ryder Street, London, for almost a month.[49] Jacques 'fasted twenty-eight days, his only sustenance being *his elixir, prepared from herbs*',[50] medicinal plants which helped him avoid starvation. Jacques's fast was later subjected to medical experiment by nutrition physiologist Noel Paton and professor of *materia medica* and therapeutics Ralph Stockman, but the 'real' composition of his elixir – as in Succi's case – was never fully clarified.[51] Jacques also ingested water, some mineral waters, and a small glass of the mysterious liquor, which he believed had the property to maintain the body weight without ingesting food for an undefined period. The press attributed an Indian origin to his elixir, but Jacques insisted that he had inherited it from his grandmother's medicinal plant recipe.[52]

Monin and Maréchal, Merlatti's doctors, described the case of Jacques as another tacit contest – in this case related to the ingestion of elixirs – because of the resistance to inanition, but stressed Merlatti's virtue for not taking any liquor during his fast, just filtered water.[53] Aiming to discredit Succi during his contest with Merlatti, Monin pointed out again the uncertain composition of the liquor, and suggested that it was probably made with cocaine, morphine, opium, caffeine, kola, maté, arsenic, and more.[54] Monin added that Tanner had used a maté infusion for his fast in 1880.[55] In his defence of Merlatti's fast, Monin discredited Succi's liquor, saying that: 'The famous liquor is just a myth, to entertain onlookers, or as a basis for further advertising.'[56]

In 1888, during Succi's stay in Luciani's laboratory in Florence, the physiologist was keen to attempt to clarify the composition and uses of the liquor in his personal interview with the faster, but, as the dialogue between the two men shows, the results were rather fuzzy:

LUCIANI: I believe that you cannot ignore the following point, if your judgement is sound and if you consider the advice of the sciences. What need is there, for example, to let people believe that your resistance to fasting is due to the famous liquor, when the merit is all yours?

SUCCI: Look, I've never claimed that the phenomenon depends exclusively on my liquor. If the newspapers have said so, I won't meddle. I will tell you frankly that the liquor is of no use to me other than to spare me the stomach pains in the first days of fasting. The main agent in my experiences is a more powerful force. [...]

LUCIANI: But in short, what do you wish to demonstrate with the fast that you intend to repeat?

SUCCI: I propose two questions to you scientists ...

LUCIANI: Queries ...

SUCCI: Two questions, and they are the following: *How do you explain my ability to combine poison with matter, to ensure that the poison becomes homogeneous with it? How can it be explained that I am able to keep my thoughts united with my body, and to maintain my strength without digestive nutrition?* [...]

LUCIANI: They are difficult problems ... it cannot be denied. But I want to hope that you will not refuse to put us on the path to resolving them. Say, would you have a hard time supplying me with a certain dose of your poison which I imagine is represented by the liquor ... ?

SUCCI: No difficulty whatsoever. But I warn you that if you wish to test its power on dogs, you won't reach any conclusions, as they tolerate even strong doses very well.

LUCIANI: Oh, is that so? ... Then I will save myself the trouble of testing it.

SUCCI: But don't think that my tolerance is just that. In Cairo I carried out some astounding tests of strength. I was good to take such a quantity of morphine and laudanum all at once that my friends were dismayed, and they were convinced that I would die. Also in Paris, as shown in a document that I will show you ...

LUCIANI: Look: the most persuasive proof you could give me of your tolerance to poisons would be to repeat an experiment with us.

SUCCI: Now is not the best time for me, dear Professor. I want to limit myself on this occasion to a simple thirty-day fast. As for the question of poisons ... I reserve the right to deal with it on another occasion.

LUCIANI: It would seem that you are putting too much meat on the fire now.

SUCCI: That's it, precisely.[57]

In Florence, the composition of Succi's elixir became a source of bitter controversy between Drs Luigi Luciani and Angiolo Filippi. Filippi led the supervisory medical board of Succi's fast and was an influential figure in that Florentine medical circle. He was reluctant to accept the standard account of Succi's elixir and questioned its composition. Filippi concluded that the liquor contained chloroform, cannabis, and morphine, and that those alkaloids remained for some time in the body before being assimilated, so they provided Succi with an extra capacity of resistance after he had ingested the drops of elixir in the first days. Although Luciani accepted the presence of these alkaloids in Succi's liquor, he considered that they did not play a key role during Succi's fast, since the hunger artist's resistance could be explained, as discussed earlier, by his special physical and psychical condition, by the strength of his nervous system and his willpower. What Filippi considered a 'physical–toxic' experiment was for Luciani a standard practice in the realm of his physiology.[58] Frequent controversies such as Luciani's and Filippi's did not help to reach a consensus on the results of the analysis of the liquor. Again the 'scientific' explanation of the causes of the resistance in such a prolonged fast remained vague and elusive.

In 1888, in Barcelona, the homeopathic doctors, and Badía in particular, described Succi's elixir as: 'an opiate liquor flavoured with chloroform and some anti-spasmodic, of which the board keeps a quantity to analyse, which he drinks to numb the sensation of hunger in the first few days ... Then he takes a small quantity of alkaline water to neutralise the stomach acids ... Can this alone suffice him to fast for 30 days or more as he does? We think not ...'[59] Badía opened the door again to hypnotism, auto-suggestion and other possible psychological factors to explain Succi's resistance to inanition (see Chapter 4) and admitted that his homeopathic colleagues did not have a convincing explanation. In 1890, once in London, before his fast at the Royal Aquarium, Succi said that his elixir was a preparation of laudanum to soothe the stomach when pain first sets in.[60] Curiously, this statement differed substantially from Succi's advertisement campaign in Florence, just two years earlier, in which his liquor was supposedly able to fight against a long list of illnesses. After the fast, the *British Medical Journal* endorsed the idea that the liquor 'resembled chlorodyne in composition and effect, with the addition of a bitter, probably gentian and capsicum, but nothing of a nutritious character has been detected, nor any coca, nor does Succi assert that he takes it for any other purpose than that mentioned'.[61] In New York, in 1890, the *New York Herald* assured its readers that Succi had drunk 710 drops of his elixir during the 45 days of his fast – not in the first days alone. Again, differing from other analyses, the newspaper suggested that it was composed of a small amount of morphine, chloroform, ether, hashish, and alcohol.[62]

In addition to these controversies, there were frequent disagreements between contemporary doctors. In 1892, for example, Neapolitan doctors discussed the issue. Antonio Carandelli, Professor of Medical Clinics at the University of Naples, noted the utility of Succi's liquor against spasmodic pain in the stomach and bowels, and regularly prescribed it to his patients. In the same university, Dr Domenico Capozzi experienced the liquor four to five times and certified its utility as a sedative for stomach pain and diarrhoea in acute intestinal catarrh. Giuseppe Longobardi, assistant surgeon at the Jesus and Mary Clinical Hospital, also in Naples, found 'Liquore Succi' very useful in cases of chronic dysentery when other drugs were not available.[63] However, these doctors did not participate in debates on the chemical composition of the liquor, which in that context seemed far from their own expertise.[64]

In practice, Succi's liquor became just another charlatan commodity, with a particular marketing campaign tinged with more or less fictional African adventures and hunger heroism.[65] The complex mixture of natural products that composed Succi's liquor and other elixir and patent medicines made it particularly difficult to achieve scientific consensus about its composition. But that scientific uncertainty facilitated marketing campaigns along with charlatan-like strategies. The tension between a liquor as a drug versus a liquor as a nutritious substance made things even more complicated, to the point that the elixir also entered academic circles, and was accepted as an intrinsic part of the hunger artists' ritual. Liquors – and Succi's elixir in particular – appeared in the more or less convincing medical reports on the causes of the mysterious resistance to the absence of food intake.

In addition to the cases of Succi and Jacques, there were other examples in which hunger artists 'played' with their mysterious liquors as part of their performances, marketing strategies, and scientific challenges for their contemporary witnesses. In 1901, for example, Alberto Magno fasted for twenty days in the centre of Buenos Aires, again supervised by a medical board; his results were published daily in the newspaper *La Capital*. Magno drank the elixir Fernet Branca, a herbal extract with digestive properties, made from herbs such as saffron, rhubarb, cardamom, myrrh, chamomile, aloe, and gentian root, in a base of distilled grape spirits. Exactly at the time of great success of public fasting, Italian émigrés had brought Fernet Branca to Argentina.[66] During his fast, also following a general pattern to prove his strong psychical and physical condition, Magno rode a bike, performed exercises, and read poetry in public. At the end, to celebrate his twenty-day fast, a theatre company performed in his honour at the Teatro Olimpo in Rosario.[67] Along the same lines, in 1905, in Vienna, hunger artist Riccardo Sacco also drank from a 'small flask filled with some secret preparation'.[68] Equally, Giuseppe Sacco-Homann – one of his competitors, with a similar name – made himself stronger by drinking Dr Adolf Hommel's Haematogen, a quack liquor that supposedly acted as a blood-forming tonic.[69] Charlatan-like elixirs therefore became an intrinsic aspect of the theatricality of public fasting.

Waters

In the satirical journal *La Caricature*, Albert Robida bitterly presented a 'devastating' landscape in which all reputable beverages would be replaced with the mineral waters that hunger artists had made popular (Figure 5.2). In a caption that illustrated all sorts of bottles, titled 'Les liquides', Robida regretted that the 'Glories of the cellars of yesteryear, triumphant Bordeaux, proud Burgundy, sparkling Champagne, Clos de Vougeot, Moulin à Vent, Pomard, Chambertin, Asti and Malaga, sweet Muscat, Argenteuil the noble name: what will become of you? The unique Hunyadi Janos [a mineral water] will replace you.'[70] Robida's exaggerated picture reflected very well the significant role played in fasting by mineral waters from different geographical origins. Like elixirs, they accompanied hunger artists during their performances, but became another source of controversy. Full agreement on their chemical and biological composition, as well as on their therapeutic properties, was hard to achieve. Moreover, the exact role of waters in the physiology and metabolism during a human fast was also controversial. In spite of all these uncertainties, doctors but also hunger artists played key roles in the marketing campaigns for the companies that sold mineral waters from natural springs in bottles.

Throughout the nineteenth century, there was little connection either between physiology and hydrology in academic terms, or between analytical chemistry and clinical therapeutics. It was often the case that the similar chemical composition of a mineral water produced different therapeutic effects in the human body.[71] Could, therefore, mineral waters be considered as efficient medicines? Was it possible to establish a direct correlation between the specific composition of a water (chemical and biological) and the efficient treatment of a specific illness, or perhaps mineral waters just acted as generalist tonics, in the same way that liquors and elixirs apparently cured a huge range of diseases? What was, then, the role of doctors in spas in terms of the medical supervision of treatments with mineral waters? Did they have to face tensions and contradictions similar to those involved in the medical boards of public

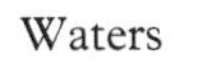

Figure 5.2. Albert Robida's satirical representation of the consequences of the 'fasting mania'. The most reputable wines, beers, and champagnes were pushed aside in favour of mineral waters such as Hunyadi Janos, associated with hunger artists' fasts.

fasts? If mineral water stimulated secretions and excretions in the body, were they relevant agents of human metabolism, and therefore linked to the physiology of nutrition and inanition? And, finally, did mineral water have a nutritious character which could shed light on processes of inanition? Whatever the answers to these questions were, they contributed in practice to the recognition and legitimation of public fasts.

During Succi's fast in Paris in 1886, Dr Marcel-Eugène Gley, a renowned physiologist of the nervous system, discussed Succi's benefits of ingesting mineral water. In his view, by absorbing a certain quantity of water rich in saline matter, such as Vichy and Hunyadi Janos, Succi protected himself from demineralisation. Gley stressed that water made up two-thirds of our body's composition. It was slowly eliminated through the skin, the digestive and respiratory mucous membranes, the kidneys, and various other glands, and it must be constantly replaced.[72] Gley pointed out the crucial importance of mineral salts for the nervous system and provided standard figures of the daily water ingestion required. Focusing more on the process of inanition, in his 1888 treatise on stomach hygiene Monin pointed out that individuals could resist starvation and autophagy for a very long time, provided they are allowed to drink water at will. He considered that water was the nutritious medium par excellence, the balancing agent of assimilation, the weighting agent of the organic exchanges that constitute life and maintain the integrity of health.[73]

In fact, the nutritious nature of water was a contested terrain in the times of hunger artists, and a serious challenge for medical and chemical experts. More generally, the controversy also reached the nature of mineral water itself as a potential therapeutic agent and the criteria for its classification and the scientific study of its action in the human body.[74] The old therapeutic classification in spas did not match well with the chemical analysis of the mineral salts dissolved in water, using the laboratory procedures of inorganic chemistry, or later with the biological, bacteriological analysis of the 1880s. To know what was in the water included looking for inorganic salts, organic matter, living or dead organisms, and the number of bacteria and their different species.[75] At the same time, these were crucial issues for the implementation of public health policies in urban and industrial contexts.[76] Fierce competition to conquer the mineral water market was constant and often involved a plurality of actors: spa owners, entrepreneurs, customers, patients, visitors, politicians, doctors, and chemists, usually with opposing views. Again, that cacophony of voices, commercial interests, and struggles for scientific authority provided a useful environment for prolonged fasting to become a lucrative commodity. Hunger artists transformed the water they drank during inanition into a key pillar of their commercial strategy.

They ascribed their names to different brands and associated their own experiences with those of the doctors who usually endorsed a specific water in advertising campaigns. Again, they wisely played on the fringes of scientific uncertainty. The traditional lack of regulation of spas and the limitations of 'scientific' hydrology left room for public speculation about the role of liquids they ingested during their fasts.[77]

During the Succi–Merlatti contest in Paris in 1886, the water ingested became a key issue of dispute among the two men and their medical boards. Merlatti only drank pure, distilled water, but Succi ingested Hunyadi Janos, which had some purgative effects and contained sodium sulphate and magnesium sulphate. Monin considered that the water Succi ingested during his fast protected him from a potential 'mineral inanition'.[78] Along similar lines, in his *Fisiologia del digiuno*, Luciani defended his thesis that sodium chloride, and more generally the salts used in ordinary nutrition, were more than simple agents of material exchange, but *true aliments*, since they were chemically linked to human tissues.[79] In addition, around 1900, the influence of water ingestion on the distribution of urinary nitrogen in animals and humans was an appealing subject of research in numerous physiology laboratories such as von Voit's and Rubner's.[80]

Hunger artists drank several brands of mineral water, with different compositions, medical properties, and geographical origin. During his fast in Milan in 1886, Succi drank Acqua di Fiuggi. From a spring in the Italian Ernici mountains, it had depurative effects on the body and was recommended for cases of kidney stones and urine and gastrointestinal infections.[81] In 1901, in Buenos Aires, Magno drank Acqua Nocera Umbra, from Perugia in Italy,[82] which rose in the Umbrian mountains and later evolved into the Bagni di Nocera, a popular spa in the region. Vichy mineral water was very common in the era of our hunger artists and many of them drank it. In the 1860s, Vichy-les-Bains was one of the most popular spas in Europe, welcoming about 20,000 visitors a year. Owned by Andreas Saxlehner in the early 1860s, the Hunyadi Janos mineral water company in Budapest, Hungary, sold the water from a cold mineral spring (today, the Saxlehner Springs). Its name, Hunyadi Janos, referred to a fifteenth-century Hungarian hero. By the end of the nineteenth century, Saxlehner had sold more than six million bottles of the mineral water worldwide.[83] The water supposedly had useful purgative properties to combat constipation, but was also recommended in cases of dyspepsia, torpidity of the liver, and haemorrhoids, as well as for all kindred aliments resulting from indiscretion in diet.[84] In 1888, in Barcelona, for example, Succi drank Vichy and Hunyadi Janos as a daily stomach wash.[85]

Figure 5.3. Riccardo Sacco, the 'Champion Fasting Man', advertising Teinach's mineral water.

The list of mineral water brands associated with hunger artists was even larger, and has not yet been fully scrutinised. In his fast in Barcelona, Succi also drank Rubinat Llorach, a medicinal, purgative water, from a spring in Lleida, Catalonia, which was particularly recommended for abdominal illnesses and gastric fevers. As a common strategy, its advertisement made use of the honours and medals it had been awarded at several international exhibitions, as well as the approval of several medical authorities.[86] In 1890, during his fast in New York, Succi drank Croton mineral water. From the mid-nineteenth century onwards, this water was extracted from the Croton river through the Croton Aqueduct, and it became the first water supply system in New York City.[87] In 1905, faster Riccardo Sacco drank Teinach's Mineral Water and appeared in its advertising campaigns, which presented him as a 'champion fasting man' (Figure 5.3).[88] Equally, Becker's Sauerstoffwasser Ozonin was one of Succi's key advertising targets during his fasts.[89] In 1915, a postcard was used as a popular advertisement with the following message: 'Sgr G. Succi only enjoys Becker's oxygenated water "Ozonin". It is the most popular and healthy table and soft drink today. Enzige Spezialfabrik Carl Becker, Hanover, Hallerstrasse, 12, telephone number 4465. Becker's oxygenated water "Ozonin" is available in all fine restaurants, hotels, cafés and forestry establishments' (Figure 5.4).[90]

The case of Agua Cruz Roja Tehuacán – known for its ability to neutralise stomach acid during digestion – and its close relation with Succi during his fast in Mexico in 1905, provides interesting hints about the feedback between the artist and the water. During his stay in Mexico City, once sealed in a cage, Succi kept some bottles of Cruz Roja Tehuacán water – a business run by Dr Lucindo Carriles – and some flasks of his famous elixir, which accompanied him on all journeys.[91] Succi's picture and a letter of him addressed to the readers appeared in a full-page advertisement in the daily newspaper *El Imparcial* with the following text: 'Better than all table water. Unsurpassed as medicinal water. Thousands cured, thousands being cured. The only ones in the world that destroy liver and kidney stones and cure stomach diseases … Each bottle bears the official stamp of the City Council of Tehuacán, as a guarantee of legitimacy. Recommended by the most renowned doctors and clinics in the Republic' (see Figure 5.5).[92]

Succi himself acted as the advertiser of the water. In a letter he sent to Dr Carriles, on 31 May 1905, Succi pointed out that:

> I can in all conscience certify that the mineral water 'Cruz Roja' from Tehuacán is very pure, light and pleasant to the palate and can compete with the best foreign mineral water, as certified in the enclosed medical certificate. That 'Cruz Roja' water from Tehuacán is truly pure, light and immensely therapeutic for the stomach is proven by my experiment of thirty days of fasting, whereby drinking this water, my body did not feel any discomfort during the whole period of starvation. I must therefore say that 'Cruz Roja' mineral water is a very medicinal alkaline water for the stomach and very effective against liver and kidney stones. If I recommend the 'Cruz Roja' water from Tehuacán, I am doing nothing more than fulfilling a duty of humanity for all the sick who suffer from stomach, liver, spleen, and kidney problems. Yours most affectionately, signed G. Succi.[93]

In the advertisement in the Mexican newspaper *El Imparcial*, Succi occupied the same status as a professional physician, Dr Fortunato Hernández, in terms of the authority to legitimise the therapeutic value of the mineral water. Dr Hernández recommended the 'Cruz Roja' brand of mineral water for kidney, liver, stomach, and intestinal diseases,[94] but, as an advertising agent, Succi also reported some of his personal feelings during the fast in Mexico and the positive effect of the Cruz Roja water on his performance. He even described how, after fifteen days of fasting, his wellbeing was increasing as he lost weight.[95]

Figure 5.4. Advertisement postcard of Becker's Sauerstoffwasser 'Ozonin' (1915).

Figure 5.5. Giovanni Succi advertising Cruz Roja Tehuacán water in Mexico.

At the centre of all these media appearances and controversies, the different types of water ingested during fasts gave hunger artists more public visibility. The narcotic/nutrition debate around Succi's liquor could be therefore extended to mineral waters. As some doctors endorsed advertisements of patent medicines in the daily press at the same time as they promoted hunger artists for the sake of their own public visibility and economic gain, hunger artists themselves took on an active role in these commercial strategies. They were often presented in the public sphere as the most extraordinary results of the ingestion of a particular mineral water, again in the tradition of the therapeutic virtues of water, elixirs, and fasting.

***

As professional fasters became just another commodity to be sold in urban marketplaces on a global scale, their elixirs and the mineral or distilled water they drank were important objects of trade, business, and advertisement. All these controversial liquors and liquids epitomised the 'liquid' nature of public fasts and the enormous difficulties that contemporary witnesses had to agree on a common description of the facts. This elixir culture placed hunger artists again in an uncertain terrain, which fitted well in eclectic medical contexts. Uncertainties about the chemical composition and medical properties of homeopathic drugs, mysterious elixirs, nutritious substances, charlatans' liquors, contemporary well-established medicines, and mineral water probably weakened contemporary scientific expertise. However, it was again in this confused terrain that hunger artists, doctors, and tradespeople gained public visibility. Homeopathic physicians used hunger artists to reinforce their own eclectic doctrines and to fight strict materialistic approaches. Confusion around Succi's elixir placed him on the edges of deceit, but also in the company of numerous doctors who, in their attempts to 'rationalise' the uses of the liquor, made Succi and hunger artists much more visible in the public arena and strengthened their professional reputation and social prestige.

At the intersection of the spa medical culture and the charlatans' market of liquors to numb the stomach, hunger artists contributed to the circulation of elixirs, mineral waters, and a huge range of drugs, which were advertised in medical journals, popular magazines, and newspapers, and became familiar commodities. In that case, cacophonies of the uncertain 'scientific truth' became an intrinsic part of the show and a rhetorical device of *the land of the hunger artists*. The potential nutritious properties of elixirs and mineral water ingested during fasts brought contemporary witnesses again to the fringes of potential fraud, but stimulated debate on the metabolic role of liquids in the processes of absence of food intake. Controversial medicines, including homeopathic drugs, were therefore compatible, even useful, for the public representation of hunger between science and hype.

Dr Benavent's homeopathic business in his pharmacy in Barcelona; Succi's commercialisation of his elixir in Florence – despite Dr Filippi's medical reluctance; Sacco's advertisement of Teinach's Mineral Water; and, again, Succi's endorsement of the Mexican 'Cruz Roja' water with a status equivalent to a medical doctor: all were key pillars of an elixir culture that shaped hunger artists and their land.

6 Politics

In 1886, an *annus mirabilis* in the land of the hunger artists, the French playwright Émile Bergerat (known as Caliban) (1845–1923) declared in *Le Figaro* that: 'The word of experimental science has been pronounced: *one can live on hunger.* The astonished universities agree on recognising this … For *socialists*, it is the extinction of pauperism. For *humanitarians* it is the abolition of anthropophagy … For [Charles] *Monselet* it is the loss of a speciality. For *workers* it is an eternal Monday.'[1] Caliban's article described the different contemporary political readings on the art of prolonged fasting. First, despite its ambivalent position, the scientific community played its cards for the sake of its authority and social prestige, since these controversial displays, in practice, gave greater visibility to doctors in the medical market. Their authority also included attempts at quantifying calories and food intake figures (see Chapter 3), in order to 'rationalise' the processes of inanition. Secondly, socialist and other leftist ideologies were concerned about the living conditions of the working class in industrial society. They dreamed of an easy solution to the malnutrition and poverty of the masses, and perceived hunger artists as potential – although utopian – social levellers. They ironically played with the potential utility of professional fasters, provided that the public accepted the nutritional virtues of controversial elixirs to reduce the food intake requirements of the population and the lower classes. Thirdly, for all those concerned with the dramatic circumstances of hunger and inanition in accidents, shipwrecks, and prisons, hunger artists provided a useful tool for feelings of charity, thirst for justice, hope to release human suffering, and trust in the body's capacity to consume tissue during prolonged periods of absence of food. Fourthly, for those specialising in gastronomy and fond of bourgeois urban pleasures, professional fasters represented a serious threat to a hedonistic, luxurious world, which they were keen to preserve. Finally, what socialists perceived as a utopian dream of a better way to feed the population, and a tool to fight for social justice, Caliban ironically described as a nightmare for the workers, condemned to probably more labour, now under even more unacceptable conditions.

Caliban's reflections can act as a source of inspiration for the political analysis of hunger artists. From the elitist reactions against the show to more or less utopian appropriations from below, the public display of emaciated bodies during prolonged fasting was linked to contemporary political issues such as discipline of the individual and society, poverty, inequality, the living conditions of the working class, the social tensions of the last decades of the nineteenth century and the cultural mechanisms to tame them. They acted as symbolic representations of political and moral values, as another – perhaps rather marginal, but relevant – weapon in the constant struggle for the cultural hegemony of the time. It is therefore the political weight of hunger artists' theatricality that is discussed in the following sections of this chapter.

In his acclaimed book, *The Shows of London*, Richard Altick regarded curiosity and spectacle as a social ethos that tamed political tensions. In his own words: 'At all times, curiosity was a great leveller.'[2] He also stressed that, overcoming conventional barriers, exhibitions of all sorts attracted lower ranks as well as the cultivated. Altick's theory was not far from Tony Bennett's reading of exhibitionary practices (museum, performances, and displays) as rituals that weave disciplinary and power relations, as a set of cultural technologies that shaped a self-regulating citizenry.[3] Inspired by the concept of cultural hegemony developed by the Italian political thinker Antonio Gramsci in the 1920s during his prison years under Mussolini's fascist rule,[4] and borrowing Foucault's reflections on mechanisms of power and social control, Bennett analysed how, behind entertainment, curiosity, and fun, museums and, more broadly, exhibitionary practices became symbols of power.[5] They legitimised the economic and political elite's discourses of progress, promoted obedience and self-discipline in order to strengthen social control, and even triggered feelings of national identity. This *exhibitionary complex* was part of a bourgeois hegemony that prioritised a specific world view in Western industrial societies.[6] It became a soft, symbolic power, a subtle cultural mechanism that put in place a set of behavioural and moral standards, particularly addressed to the lower and middle classes.

In their time, hunger artists combined Bennett's three levels of the exhibitionary complex: discipline, surveillance, and spectacle. Their itinerant journeys led them to places of discipline such as the *asylum* – being resistant to inanition often placed them on the fringes of illness and insanity; the *clinic* and the laboratory as places of systematic experiments in which they often stayed in sealed cages such as respiration calorimeters; the *prison* – which was always a threat for their intrinsic charlatan, fraudulent nature as simple peddlers selling miraculous medicines, and the necessary surveillance and medical supervision of their fasts; and the *museum* (including theatres, industrial exhibitions, and world's fairs) – as the multiple sites in which hunger artists encountered the general public and displayed their physical and intellectual skills through sophisticated performances (see Chapter 2). In fact, hunger artists, and all the historical actors who accompanied them during their performances, constructed metaphors of self-discipline, which could be extrapolated to social discipline and power. They also shaped public health issues such as debates on nutritional standards for the working classes, and acted as bridges between the academic power of the medical elite and the huge terrain of eclectic, popular medicine, in the struggle for scientific authority and social recognition. In addition, despite their itinerant and cosmopolitan aura, hunger artists also contributed towards strengthening the narratives of the making of an ideal citizen as a key actor to construct the modern nation-state.

Just as mid-nineteenth-century industrial workers travelled from northern England to London to visit the fascinating steam engines and machines of the international exhibitions, and went back home accepting their structural role in the tough everyday life of the factory,[7] hunger artists, exhibited in industrial cities, could stress issues of discipline, reflections around the minimum food intake needed and issues about overfeeding of the upper classes, and trigger debates on optimal feeding of the average citizen of the nation. Although their displays were probably more modest than the ambitious performances in international exhibitions and museums, hunger artists can be considered as another exhibitionary complex, as semiotic devices that unveil issues of modernity, progress, popular culture, and national identity. Beyond disciplinary boundaries and professional barriers, nineteenth-century spaces of food preparation, consumption – and abstinence – were also spaces of knowledge making, which, in practice, became places of power.[8] As commodities in the marketplace, far from the old religious, mystical tradition of asceticism and personal mortification as a way to access the divine sphere, public spectacles of prolonged fasts brought to the fore issues of food consumption and abundance and its rational use for the sake of a healthy individual and a sane society. Often in a theatrical manner, the therapeutic uses of fasting for medical purposes soon leaked over into the therapeutics of society.

As discussed in earlier chapters, in his cartoons, Robida ridiculed the contemporary fasting mania, and praised the bourgeois pleasures of French gastronomy. His comments were, though, intriguing: 'Mr Succi therefore works, without perhaps even thinking about it, on solving the social issue in favour of the poor against the rich.'[9] What were Robida's political intentions in that sentence, in the ironic context of *La Caricature*?

Were Succi and other hunger artists social levellers? The Robin Hoods of the poor, of the lower classes, of the industrial workers, who could now live with less food? Or perhaps hunger artists were just charlatans who played their cards in a global marketplace with no explicit political position? An appropriate answer to these questions would probably unveil the very political nature of the performances of the professional fasters. In this land of the hunger artists, contemporary actors' behaviour helps us to easily cross the boundaries between the natural, social, and political fields.[10]

In 1890, Dr Henri Labbé, summarising Richet's lecture on inanition at the Sorbonne, commented on the social and political dimension of hunger: 'in the Parisian population, decay, disease, tuberculosis ... were the result of a deplorable diet ... *The matter of starvation, extended to that of insufficient food, particularly exceeds the physiological framework that we like to draw for it. It is a social matter*, a matter of the vitality of the future of the race, which can be resolved only by the progressive instruction and education of all.'[11] The challenge was, though, how to reconcile concerns about the insufficient feeding of the population with practices of inanition as therapeutic, social, and political issues.

Discipline

In Rudolf Virchow's view, the medicine of his time should be regarded as a social science, since he considered that politics was nothing but medicine on a large scale. In his *Report on the Typhus Outbreak of Upper Silesia* (1848), Virchow treated his patients with drugs, but also encouraged changes in food habits, housing, and clothing.[12] In fact, a healthy life – in the tradition of Alfons Labisch's *homo hygienicus* – required self-discipline and social discipline for the public benefit.[13] Hunger artists' performances can be therefore considered as another manifestation of that culture of self-discipline, in this case, as a spectacle of the representation of a human body in pain and suffering, but also, as an athletic-like achievement, an heroic act of resistance to inanition.[14] After contemplating the emaciated faces and bodies of hunger artists, visitors became aware, more or less consciously, of the values and the morality of the bourgeois industrial societies of the time. Self-responsibility, personal hygiene, exploration of the limits of the human body, and individual capacity for effort and suffering tacitly spread through the subtle codes of the professional fasters. In addition, the extraordinary resistance to inanition by professional fasters suggested the possibility of an excess food intake by the population at large and therefore it opened the door to a review of food standards.

In the late 1870s, the famous American entrepreneur George Francis Train – who had joined Succi on part of the latter's journey through the United States (see Chapter 2) and also practised abstinence from food – praised the moral, self-discipline virtues of fasting. He described his own experience in the following terms:

I have proved three things. The first is that all the stories of terrible agony in starvation are humbug. They are inventions of people who want to glorify themselves because they have an opportunity. As a matter of fact, hard times are a blessing to the country, because they cause very many persons to fast from sheer necessity. The second thing I have proved is that I, who am not a Christian, but who accidentally have all the virtues and moralities of a Christian, have really been able to do what the lying monks only pretended to do. I have really fasted, and have really felt my intelligence and power of prophecy increased. The third thing I have proved is that a person who has fasted six days has no ravenous appetite.[15]

Francis Train considered individual efforts of abstinence – now as a secular endeavour and beyond any religious mysticism – as a 'blessing for the country', as a moral value to strengthen the virtues of society. As a result, the charlatanesque, suspicious nature of hunger artists could be seen as an exemplary practice, as a collective lesson. In a similar context, this glorification of the moral virtues of fasting often intersected with other medical, social, and political reflections (Figure 6.1). In New York in 1880, Robert Alexander Gunn, at the time professor of surgery and dean of the US Medical College and member of the medical board that supervised Dr Henry Tanner's fast, intended, for example, 'to call public attention to the importance of knowing more of the subject of food, its uses and abuses'.[16] Gunn focused on the physiological study of food ingestion and inanition, but also opened the door to a broader debate on the ideal figures of food intake for average human beings, an issue that inevitably carried political weight. In this case, the therapeutic uses of fasting from a medical perspective could also be extrapolated to its 'therapeutic' uses for public and social health.

Reasons in favour of the individual and collective virtues of prolonged fasting came with ample criticism against the supposed overfeeding of the population. In 1886, Dr Henri Perrussel, a French physician and homeopath, questioned, in moral terms, the immoderate human ingestion of food. In an article on voluntary, prolonged fasting, he pointed out that: 'humans eat a lot more than they should, and therein lies … one of the main causes of the majority of their diseases, more common among them than among animals'.[17] Perrussel praised the virtues of diet and attributed the impressive longevity of the famous French chemist Michel Eugène Chevreul (1786–1889) to his long-lasting austerity in food consumption.[18] Adolphe Burggraeve (1802–1906), professor of

Figure 6.1. Giovanni Succi giving an exhibition of sword exercise and self-discipline during his fast in Barcelona in 1888.

medicine at the University of Ghent, Belgium, also endorsed the theory of the excessive ingestion of food among the population and the need for self-discipline: 'It is certain that our meals are too close together, either because of accepted customs, or because of sensuality, or because of a kind of acquired idiosyncrasy which does not allow us to resist hunger. It would be simpler, and more natural, *to have only two meals a day*, one in the morning and one in the evening (between seven and eight). This would be the social problem to be solved and would at least be of practical use, whereas our fasters only serve to feed the curiosity of onlookers.'[19] Disciplining the population with a more moderate food intake, two meals per day, was in Burggraeve's view a 'technical' solution with obvious potential social consequences. In 1887, Burggraeve published a book on human longevity,[20] which in his view was based on the virtues of *dosimetric medicine*, another heterodox, eclectic medical branch that used active principles of the chemically pure compounds – often alkaloids – to be administered as drugs in regular, well-established doses to the patient. The appropriate doses and the appropriate diet or abstinence was, in Burggraeve's view, an ideal combination for better health and longevity.

The social concerns of prestigious medical doctors on the overfeeding of the population also led to reflections on the public impact of hunger artists' performances. Dr Giovanni Chiverny, president of the medical board for Succi's fast in Milan, even considered some psychological benefits of abstaining from food: 'If everyone could get used to eating little and once a day there would be more time to cultivate the mind.'[21] In 1890, for example, after Succi's 45-day fast in London, the prestigious medical journal, *The Lancet*, said that: 'Perhaps the conclusion may be drawn from this experiment that a considerable proportion of our daily food is not applied to any useful purpose in our economy, and that many of the inactive inhabitants of cities *habitually eat more than is required* to maintain their mental and bodily functions in the highest efficiency.'[22] In his *Aperçus de Médecine sociale* (1905), neurologist Louis Landuzy (1845–1917), drew the population's attention to what he called the '*art de l'alimentation*'. Landuzy considered humans to be the animals who used their instinct of nutrition least and who made the most mistakes in their diets and regimes: 'the one who knows least about not eating badly, not drinking too much and not eating too much'.[23] Following Virchow's tradition, Landuzy has a particular sensitivity about the social causes of illness, which he exhaustively studied in the case of tuberculosis, but that also extended to human eating habits and abstinence.

Hunger artists' performances proved that resistance to inanition in humans was higher than expected, so perhaps human bodies needed less food to survive than was thought. Public fasts could therefore lead to debates on nutrition and diet and on the limits of human feeding as well as on the therapeutic virtues of fasting, and borrow them from different ideologies and political purposes. In 1892, for example, during hunger artist Alma Nelson's fast in Paris (see Chapter 4), newspapers such as the Spanish *El Heraldo de Madrid* stressed that the aim of the performance was to give to the poor the means to live without food without jeopardising their health. The whole issue turned into a moral discussion on the virtues and vices of eating:

> Do our readers remember the famous fasters Succi and Merlatti? Well, now in Paris they are going to have not a fasting man, but a fasting woman, an American artist, Miss Nelson. Like her predecessors, she will be surrounded by a board of doctors who will watch over her day and night, and ensure the truth of her statements. Miss Nelson forces herself to drink no more than a glass of a special drink every day. *Miss Nelson's aim is to give the poor the means of going without food without compromising their health.* She will begin her experiments today, Sunday, at the Grand Hôtel de Paris, and during her fast she will perform four concerts for the benefit of the poor of the French capital.[24]

Amidst debates on diet, abstinence, and the therapeutic value of fasting, hunger artists also reinforced vegetarianism. Some contemporary reactions to the 1886 Succi–Merlatti contest in Paris endorsed food moderation, diet, vegetarianism, and abstinence in the following terms: 'Cooking has killed more men than war ... it is not an irony of Providence that these experiments of Succi, Merlatti, and others who have come to demonstrate to the men of our time, who aspire only to sensual enjoyment, that the body can be subjected, without losing too much, to repeated, sustained, even abusive fasting; that this society of vegetarians is going to prove – by practising it openly – that vegetarian food can make man live as well as, and more normally even, than carnivorous food.'[25]

Some years later, in 1892, a controversy between Charles Richet and the Russian writer Leo Tolstoy on '*l'alimentation*' appeared in the journal *Revue scientifique*. Tolstoy defended the moral condition of fasting and abstinence, in general, to reach sobriety and control human passions. In fact, he considered that, as a training exercise, a relative fast, one which avoided animal food, could contribute towards achieving that moral sobriety. After a very touching literary description of animal suffering in Russia, Tolstoy defended vegetarianism and refrained from eating meat, since he considered that it was intimately linked to the immorality of assassinating animals, provoked by gluttony.[26] The vegetarian morality and the virtues of moderate fasting paved the way towards human perfection and a path to make the Kingdom of God true on Earth. Tolstoy also remarked that it was in Germany, England and America – not in France – where the traveller could find vegetarian hotels and restaurants. In his words: 'the first condition of a moral life is abstinence, the first condition of abstinence is fasting'.[27] Tolstoy defended three main theses, which had important analogies with debates surrounding the morality of hunger artists: (1) Luxury is bad; (2) We eat too much; and (3) Animal food should be replaced with vegetables.[28]

Nevertheless, Richet, a key figure in the scientific study of animal and human inanition, perceived vegetarianism as an outmoded return to nature and defended scientific, industrial progress, which included 'luxury and wellbeing' in a refined civilisation. It was therefore up to the scientific authority of physiology (and physiologists) – not to religion – to decide the required minimum food intake for humans. Richet mentioned a daily standard of 125 g of meat, 300 g of bread, 300 g of potatoes, and 50 g of butter and cheese. Beyond these limits, which included meat, there was a risk for individual health.[29] Richet concluded that 'there may be some advantage to man in eating meat, although meat is not essential. This is the opinion adopted, with reason, we believe, by most physiologists.'[30] The floor was opened to huge debate across national borders.

Was it possible, for instance, to fulfil Carl von Voit's protein standard for German workers (118 g/day) without ingesting meat? Were vegetarianism and fasting therapeutic practices, as many doctors and social reformers believed throughout the second half of the nineteenth century? Nathan Zuntz, who in 1893 had studied hunger artist Breithaupt in his laboratory, considered for example that an average of 80 g of protein per day could be met with vegetables alone.[31]

In their habits, controversies, and performances, even in their silence, vegetarians and diet reformers, but also professional fasters, played a civic role. The latter, as celebrities of strict, disciplined habits to resist the absence of food, had a double political dimension. On the one hand, they empowered individuals to improve their health and self-control, as part of the hygiene and public health campaigns of the time; on the other hand, they theatrically represented the overfeeding of the population as a general statement that subtly circumvented the painful gap between the opulent banquets of the upper class and the food scarcity of the lower classes, who were provocatively encouraged to revisit their food intake standards and to agree to eat less.

Poverty

Charles Richet perceived fasters such as Tanner, Succi, and Merlatti as cases of privileged men with shelter, food, and security in their future, compared to the social pain of fasts caused by mining accidents, prison, shipwrecks, and poverty.[32] In bourgeois urban contexts, hunger artists acted therefore as a warning of the extreme, miserable conditions that human beings could potentially face. They could even jeopardise the comfortable social conditions of the urban upper classes. In practice, though, concerns about the miserable feeding conditions of the lower classes were part of the agenda of the scientific and political elites of the time. The fasters' exhibitionary complex (à la Bennett) reinforced a specific bourgeois hegemony, which did not attempt to substantially alter the social order, but just to speculate on the food intake of the lower classes. Robida's caricatures of the Parisian restaurants attacked by the *jeûnomanie* were just part of the same play.

Hunger artists' cases also contributed towards bringing nutritional issues to the realm of the popular and to opening public debates on hunger, poverty, and social injustice. It is worth remembering here that Tanner introduced himself as a firm friend of 'liberal and progressive medicine', a follower of American eclectic medicine, a more 'popular' version, which praised Native American roots.[33] The old resonances of Tanner's fast and his defence of 'popular' medicine integrated several controversial practices and can be therefore seen as a seminal movement of future fruitful developments in the marriage between hunger artists and the popular realm. In 1880, although the US Medical College supported Tanner's daring initiative of fasting in public for forty days as an interesting scientific experiment, members of the New York Neurological Society, Dr William Hammond among them, bitterly criticised the fast as fraudulent and unprofessional in terms of their own medical culture – a pattern that was later reproduced in Europe. In spite of the controversies, the move towards the popular and its links to fasting had deep roots in American eclectic medicine, and had its continuities with other later advocates of natural therapies. These included physician Dr Edward H. Dewey (1837–1904), author of the controversial and widely read *The No-Breakfast Plan and the Fasting Cure* (1900), and his pupil, Linda Hazzard (1867–1938), who in 1908 published *Fasting for the Cure of Disease*; both became great defenders of the medical and moral virtues of diet and abstinence (see Chapter 4). Dewey, like several doctors who studied and followed hunger artists, held a rather marginal position in terms of academic authority, but his popular books achieved wide readership.[34]

In France, others also linked hunger artists' performances to claims in favour of 'popular' science and medicine. Science populariser Victor Meunier, for example, was a member of the supervisory board of Merlatti's fast in Paris, and reported details of the Succi–Merlatti contest in his newspaper *Le Rappel*. Because of his socialist, leftist ideology, Meunier was against orthodox academic science and in favour of popular knowledge which would strengthen the intellectual autonomy of the masses. In his *Scènes et types du monde savant* (1889), Meunier denounced bureaucracy and the weaknesses of French academic science and its defeated position compared with Germany and Britain. He perceived academic science as 'a science of civil servants, and civil servants of one and the same administration, where emulation, noble rivalry, and free competition are, like independence, as unknown and impossible as great wingbeats in a cage. This is the evil from which we could die. We are heading for scientific bureaucracy.'[35] Meunier supported young doctors and scientists – such as Monin and Maréchal, who like Meunier were on the medical board of Merlatti's fast – to fight for their professional careers in that rigid elitist system but also to struggle for a social medicine to combat poverty. Meunier praised popular knowledge as a useful tool to empower the lower classes.[36] In that framework, he regarded hunger artists, their controversial image, and the plurality of voices that appropriated them as useful tools to fight against the monolithic, elitist science of his time.

Figure 6.2. Ernest Monin's *L'hygiène du travail*, front page (1889).

In a similar vein, Dr Ernest Monin, one of the promoters of Merlatti's fast in Paris, in 1886, was a member of the Société française d'hygiène and was known for his sensitivity towards public health and popular medicine (Figure 6.2). Monin published extensively on issues such as obesity, eating, practical hygiene, digestion, feeding, alcoholism, labour, sex, the stomach, diabetes, and beauty, as well as eclectic practices such as hydrology.[37] He was also widely known for his popular medical aphorisms. In Monin's view, hunger artists were ideal bridges between expert and popular medicine, and their public impact reinforced hygiene and public health campaigns to fight hunger and poverty.[38] In his discussion on the causes of human longevity, Monin pointed out that: 'The poor die twice as much as the rich: poverty is the great provider of death. City dwellers also die twice as much as country people, precisely because poverty is more common in big cities.'[39] Equally, he advised general, moderate food intake to have a longer life: 'follow a sober diet, remembering that "the mouth" … has claimed more victims than the sword'.[40]

Hunger artists themselves often publicly expressed their own political opinions on issues related to poverty and social inequality. Succi gave a paper at a medical conference that took place in Barcelona during the 1888 International Exhibition. He warned about the risk of hunger artists providing useful arguments to the bourgeoisie to dismiss working-class demands. The conference organisers summarised his paper, as reported in the official journal of the exhibition: 'Read a communication from Mr Succi, asking the conference to appoint a board to check and audit his 30-day fast and bodily exercises (the fasting man gives arguments to the satisfied bourgeois to dismiss the claims of the miserable working classes; for they will tell them that one can live and work with little food, e.g. Succi).'[41] In Madrid, the same year, Succi's impact was rather marginal, but again raised debates on the moral meaning of hunger, in the face of urban poverty and social marginality.[42]

Bringing to the fore the social question and the inequalities between the rich and the poor, hunger artists also appealed to the moral value of Christian charity. In 1886, professional faster Mr Simon asserted that his motivation was humanitarian, to help, for example, miners trapped in a mine.[43] In 1888, in Barcelona, Succi donated 20 per cent of the income from his visitors' entrance tickets to charity.[44] In 1895, the conservative newspaper *Diario de Barcelona* considered that: 'It is for physiologists to determine the effects of starvation and to discover them in their frightful reality, for the scalpel is insensitive to compassion; yet *charity* never loses its rights, and it is responsible for alleviating human suffering … This must be the conclusion of our study; *let us multiply the charitable societies*, let us give more, let us always give to help the unfortunate in their struggle to exist, and let us do our utmost to ensure that history does not repeat the horrible dramas which unfortunately it so often records in its pages.'[45]

Inanition and hunger therefore became driving forces to raise issues of not only poverty, social inequality, and attitudes towards charity, but also the calculation of the minimum calorie intake required in terms of individuals' productivity in industrial societies. In the latter case, hunger artists became a powerful metaphor of austerity, again extrapolated from the individual to society. From Liebig's animal chemistry and the sale of his meat extract to Moleschott's mechanical explanations, the physiology of nutrition and the study of human metabolism became central in the complex process of rationalisation of diet, in which hunger artists could be used from different perspectives. From Carl von Voit's study of the role of nitrogen in food to Max Rubner's calculation of the nutritional value of calories, experts' opinions on the limits and nature of food intake and the therapeutic value of abstinence could be presented in an 'objective' way, but, in practice, scientific reports were inevitably influenced by the social and political issues at play at the time.[46]

In 1886, French populariser Wilfrid de Fonvielle had already described the '*ration d'entretien*', that is to say, 'that which is necessary to ensure that the weight of the living being neither increases nor decreases, but retains the exact value it possesses, at a time when it is neither too thin nor too fat'.[47] Fonvielle was particularly interested in the standardisation of the material exchange in human respiration and in the calculation of the average energy of human life, which he linked to hunger artists with the following reflection: 'The daily exercise that fasters perform, when they go for their walk in the Bois de Boulogne, in the streets of Turin, or around the lake of Chicago, can be accomplished only at the expense of meat and fat, which pulmonary combustion will consume.'[48] Fonvielle explained that an average man exhaled around 760 g of carbon dioxide per day, that is to say, around 200 g of carbon, and that this was the quantification of the average energy of the human life. He pointed out that the mechanical equivalent of heat taught his contemporaries that the more physical work performed, the higher the carbon consumption.[49]

In collaboration with Angelo Mosso, Nathan Zuntz worked extensively on metabolism, nutrition, respiration, and high-altitude physiology, so mountaineers and aeroplane pilots became another valuable object of physiological research.[50] As Mosso had discussed in depth, the word *fatigue* had multiple meanings at the time of hunger artists. From a mechanical, physico-chemical perspective it could refer to muscular weakness, low body performance, and changes in the chemical composition of blood. From a psychological point of view, it could be linked to mental laxity and slow reactions. In terms of industrial production, it was closely related to accidents, occupational illness, low productivity, and industrial unrest. The triple nature of fatigue, which encompassed all these physiological, psychological, and philosophical explanations, also applied to education. Referring to Jacob Moleschott's book *Lehre der Nahrungsmittel* (1858), Mosso defined intellectual fatigue in the following terms: 'In artists and scientists the material exchange promoted by their intellectual exertions is again moderated by their sedentary life. Nevertheless, we find in them as effects of their activity of mind a more abundant secretion of urinary salts, increase of bodily temperature, and a greater need of nourishment.'[51] Mosso defended a physiology to improve life in humans, but more particularly for better conditions for the working class. His master, Moleschott, as a representative of materialistic physiology, added a moral political value to issues of nutrition, diet, and inanition. He perceived nutrition as a 'science for the people', which had to be devoted to their physical wellbeing, dignity, and freedom.[52]

The assessment of fatigue could therefore become an indicator of the price paid for the industrialisation of modern societies, which inspired the physiology of work and working conditions around 1900. As in the case of hunger artists, academic expertise and social conditions shaped concepts such as industrial fatigue.[53] Like fatigue, hunger could be understood as a material, physical entity of the body, an 'objective' lack of food that could lead to discomfort, restlessness, and death, or a sort of 'appetite of the mind', a metaphor on individual desire and world views, but also a deep political object which reflected the social health of the time.[54] In the context of the science of nutrition, hunger artists therefore appeared as a useful bridge between specialities and audiences. However, the science of the balance between calories, proteins, physical activity, physiology, and economy could easily become a technocratic depoliticisation of food requirements and social justice.[55] The scientific study of nutrition could be a promise to solve hunger as a social problem but, at the same time, a weapon against working-class rights. This was the uncertain, even paradoxical, terrain in which food (and hunger) became a polysemic word appropriated from different political agendas.[56]

Although medical expertise and nutritional modernity provided a quantification of the minimum standard intake of food, calorie figures, and tables, and the apparent objectivity of ciphers, numbers, and tables, it left the social consequences of the performances of voluntary, prolonged fasting open to discussion. As the sophisticated instruments of experimental physiology faced limitations for a full, convincing explanation of the causes of human resistance to inanition, and opened the door to non-materialistic, psychological factors, the standardisation of required calorie intake could not be isolated from political and social factors.

Nations

As discussed in recent historiography on the links between science and the modern nation, the construction of a mythical past – the invention of tradition – is a key issue in the making of a national identity, but at the same time science and technology – in particular in the nineteenth century – provided excellent examples for the public accounts of modernity and progress, as a key issue to attract citizens to construct a common future.[57] In addition, nineteenth-century spectacle and exhibition, with all their glamour, reinforced the elite's programmes of social control and discipline, which often included nationalist values. Disguised as cosmopolitanism, public displays of nationalism triggered the rivalry that led to the cruel confrontation of the First World War (at the end of the period of public success of hunger artists) and reinforced Eurocentric colonial policies.[58] Although the land of the hunger artists was transnational and cosmopolitan in nature, a nowhere place in which cities, nations, networks, news, laboratories, and spectacles seemed to weave a seamless web (see Chapter 1), its public impact did not escape appropriation in specific local contexts and provided notable differences at a national level that are worth exploring here.

In France, for example, voices that were reluctant to validate hunger artists' performances, such as populariser Wilfrid de Fonvielle, considered that welcoming them in scientific academies or even discussing the scientific nature of the fasts would end in profound humiliation of the nation.[59] Fonvielle believed that the case of hunger artists, with its intrinsic charlatanesque nature, posed a threat to Republican values and the scientism of the Third Republic, as well as a humiliation for the scientists who took them as serious objects of research. In his book *Mort de faim*, he stated that: 'These fasts are part of a conspiracy against human dignity and the honour of the Patria. What joy would the monarchical states have if absurdities ... were recognised and proclaimed by the public authorities, if the claims that have dared to arise in the Academy of Medicine and the Academy of Moral Sciences were accepted by the Senate and the Chamber of Deputies of a republic that prides itself on basing its social state and its policies solely on reason.'[60]

But others perceived French national pride from a different perspective. In Paris, in a public lecture at the Conference Hall on Boulevard des Capuccines, Charles Laisant (1841–1920) praised the value of Merlatti's fast for the progress of physiology. Laisant was a mathematician, director of the journal *Le Petit Parisien*, and future president of the French Association for the Advancement of Science. From his strong nationalist position, he represented the public sentiment that France had probably lost its scientific hegemony after its defeat in the 1870 Franco-Prussian War. Laisant felt proud of the successful appearance of hunger artists in Paris, as an object of experimentation for French physiology in its rivalry with German medical schools. He hoped that this kind of event, which had a strong global impact (see Chapter 2), could put France back at the centre of scientific and technological leadership, at a time in which British inventiveness and German science-based industry seemed to have relegated the French academic system to a secondary position.[61] Although Fonvielle and Laisant clearly disagreed on the use of hunger artists for the prestige of the Republic, the two men shared deep nationalist concerns when addressing those performances.

In other national contexts, a public fast was interpreted differently by contemporary public opinion. Succi's performance in Madrid in 1888, for example, did not attract much interest in a country with high figures of famine and illiteracy. Journalist Enrique Sepúlveda's description of the event in *La vida en Madrid* (1886), a report on everyday life in the Spanish capital, offers some hints: 'The experience is strange and odd; it would be more so in another country, because here where a large part of the population lives on air … here where the bricklayers and carpenters who work on the buildings have enough with a pepper or a piece of cheese; here where the school teachers do not eat, at least as far as I know; they must not be able to eat, rather, because of the horrible intermittencies they suffer to earn their salaries; here, I repeat, the Succi fast does not surprise as much as it should.'[62] Sepúlveda wondered ironically if Succi's liquor could feed all the hungry Spanish people: 'Because if so, Succi would be a true Saviour of humanity, and would have solved a problem more arduous and more difficult than that of living without work, which would be that of living without eating … in this obscurity, in this unclarified doubt, lies the little interest that the show has raised in Madrid, and the scarcity of visitors to the Teatro Felipe.'[63]

In terms of national humiliation, a sense of incompatibility between a public prolonged fast and the country's misery also appeared during Succi's performance at the Science Pavilion of the 1888 Barcelona International Exhibition. Journalist and politician Federico Rahola (1858–1919) bitterly criticised the public exhibition of inanition in a science pavilion. He considered Succi's fast as an insult to a nation that was clearly backward in terms of science and technology. Rahola pointed out that: 'Confess that you have been mistaken; during your fast at the Palace of Science, you have been a symbol, a real abstraction. This prolonged thirty-day abstinence in the midst of educational works, of the products of scientific and literary associations, of the display of the

Figure 6.3. The Palace of Science where Giovanni Succi fasted for thirty days in Barcelona in 1888.

efforts of science and study in Spain, symbolises the absences and fasts to which the unhappy man who devotes himself to scientific speculation and teaching is condemned, dismissing bullfighting, politics, and the stock exchange. It often seems that chance has a conscience, and we have never had more reason to suspect it than the fact that your fast took place in the Palace of Science' (Figure 6.3).[64] Science and technology were heated issues for Spanish national pride and identity at the end of the nineteenth century. The 1898 war against the United States, and the loss of the last overseas colonies of Cuba and the Philippines, was perceived as a national humiliation and was associated with a national weakness in terms of scientific progress. What an insult it was, therefore, to host a charlatan in a science pavilion.[65]

As discussed earlier, in late nineteenth-century Germany, eating 'naturally', including vegetarianism, other diets, and in some cases – again – the use of fasting as a therapeutic strategy, became part of the biopolitics of the country's modernity. With a physiology of diet and fasting that coevolved with civil society, the *Lebensreform*, as a cultural movement, included vegetarians, naturopaths, nudists, diet reformers, anti-industrialists, and popular health practitioners. Like hunger artists, some *Lebensreformers* such as vegetarians became objects of scientific investigation.[66] Edward Baltzer's natural lifestyle, for example, opposed meat and trends towards industrially produced food.[67] Others also inspired a neo-Hippocratic naturopathy that promoted sunbathing, a vegetarian diet, pure air, water cures, and organic agriculture. Originally, the movement had popular roots but it progressively influenced the elites and became a feature of the national identity. Max Rubner, the prestigious physiologist who had developed one of the most sophisticated respiration calorimeters to quantify the metabolism during fasting (see Chapter 3), and contributed to the construction of the caloric vision of the human body, criticised Germans for eating too much meat.[68] In that context, hunger artists again represented a useful metaphor for therapeutic diets, food moderation, asceticism, and a more respectful way of living with animals, which could be assimilated to key aspects of the German identity. The boundaries between medicalisation and nationalisation of hunger were in this case elusive and blurred.

Italy was at the core of the national identity of the land of the hunger artists. The longstanding – since the Renaissance – Italian tradition of charlatanry and itinerant peddlers probably had something to do with the overwhelming majority of Italians among the professional fasters under study. On his trips to Africa during his youth as an agent for an Italian trade company (see Chapter 1), Succi, again, represented the colonial, nationalist Italian enterprise. He also campaigned for the defence of the virtues of his resistance to inanition for the progress of Italy, a young nation-state at that time, after its recent unification. In 1890, Succi stated that: 'If only all the subjects of our good king Humbert would follow my example, Italy might soon find a way out of her financial difficulties, abolish taxes and become a flourishing nation.'[69] Not by chance, a professional faster of German origin, Wilhelm Bode – like others – Italianised his name to Riccardo Sacco as a marketing strategy.

From the perspective of Kafka's short story, in the 1920s, hunger artists were regarded as part of a decadent profession, whose talent and artistic creativity had lost their momentum, as the impresario replaced the faster with a panther, at a time in which cinema, mechanical recreations, and other spectacles and entertainments had begun to captivate the interest of the urban masses. From that perspective, the role of hunger artists in the modernisation and progress of the nation seemed rather problematic, since they were seen not to fit into any scheme of modernisation and scientific progress, which was usually associated at that time with some of the main virtues of the making of the nation.[70] Their status referred to the old charlatan, freak show tradition, held over from early modern times. Nevertheless, at the same time, they seemed to be a relevant part of the 'progress' of experimental physiology in different nations at the end of the century, with notable contributions to the study of the human body and its metabolism in conditions of food deprivation. Although the land of the hunger artists transcended localities, the multiple uses of public fasts seldom escaped nationalistic rhetoric.

***

As Albert Robida ironically demonstrated in *La Caricature*, public fasts could become a useful experiment to revisit minimum food intake figures for a standard living – with a particular emphasis on the feeding requirements of the working class – but also a dangerous message in the urban bourgeois context of restaurants.[71] In that satirical journal, Jules Demolliens, who usually added his articles to Robida's drawings, created a fictional tale about a futuristic society, which had adopted fasting as a new religion. Demolliens imagined that, in 1960, after decades of proselytism since the events of 1886, a '*succiste*' circle, and its journal *L'Anti-restaurant*, had conquered the souls of the French population. Food professionals, an army of butchers, bakers, and grocers, unsuccessfully revolted against the *succistes*, who finally won the battle and opened the door to a new fasting civilisation. A long time later, in 2150, France had a constitution in which fasting was mandatory, and all restaurants, cafés, food shops, and markets had been irrevocably closed. Nevertheless, in 2350, new opposition against the *succistes* achieved some results, not to overturn the situation, but at least to achieve social peace for a time. Then, gold and silver ingots had no value and became bricks to pave the streets. Coins and notes were useless too. Weavers and builders had no jobs, and clothes and shelters disappeared. No more jobs, just leisure. In the countryside, fields and crops lost their natural space in favour of huge forests that conquered the periphery of Paris, while sheep and other animals reached the Seine. The cattle population decreased, since they could not find meadows in which to graze. It was a real golden age, in Demolliens's imagination. Nevertheless, hundreds of years later, in 2560, the inhabitants of Paris woke up to an unexpected spectacle. A man had just arrived, and he attempted to introduce pieces of meat, vegetables, and fruit into their mouths twice a day. There was great scepticism among the population, and suspicion of fraud. Many considered the visitor as a rare oddity, who was not to be trusted; but he obtained the legitimation of a medical board. Then he quickly achieved the acceptance of the crowds. Demolliens regretted how easily again the public could opt to eat or fast in quite a naive, alienating way.

Demolliens's tale helps us to become aware of the social and political nature of fasting. Behind the apparently innocent performance of a public fast lay a powerful exhibitionary complex, a deep political message.[72] Metaphors of emaciated bodies could be extrapolated to metaphors of society, so different medical regimes of food intake reflected the class structure, and the discipline of the body became social discipline.[73] Like hotels, theatres, monuments, exhibition pavilions, and bridges, fasting places acted like Walter Benjamin's Parisian 'arcades'; that is to say, in the same way that all those urban sites had disappeared after the

Hausmanian reforms, but had retained a considerable part of the identity of the old city along with their cultural, political meaning.[74] They were considered as useful places for rituals of curiosity and fascination of the crowds, but also as sites for social control and stability. Benjamin described in detail several cases of early elements of consumption in the city: lithography, photography, gas lighting, fashion, railroads, and industrial exhibitions. In his view, all these objects or spaces represented a sort of dreamscape of the desires of the French urban bourgeoisie, a key step towards a new commercial fin-de-siècle modernity that combined spectacle, retail, shopping as leisure, and technologies of promotion.[75] Beyond the emblematic case of Paris, this description fits very well, in my view, with hunger artists' performances and the sites through which they transited and exhibited their emaciated bodies. They occupied urban spaces to become just another icon of the bourgeois cultural hegemony of their time, between consumption, newspapers, advertising, and spectacle.[76]

In the land of the hunger artists, there was no clear boundary between food, hunger, social stability, and struggles for power and control. Voluntary hunger evoked a great variety of collective emotions, from fury and rejection to adamance and sympathy, in a struggle for the 'economy of attention'.[77] Richet paid particular attention to forced fasts, that is to say to the 'situation of individuals who have been caught, for example, under a landslide … the unfortunate ones thus separated from the rest of the world know that they cannot be helped: that in order to reach them, they have to break through passages and clear several hundred cubic metres of earth and stones. In these conditions, the deprivation of food is sometimes very long … The miners of Bos-Mouzil remained stuck for eight days following a landslide, without suffering much. The latter gave a rare example of human solidarity and charity. In general, individuals subjected to the anguish of hunger show a ferocious egoism … Other examples are cited in these forced fasts, no longer of miners, but of shipwrecks.'[78] Leaving the dramatic circumstances aside, one should conclude that Richet's three different kinds of fast – experimental, charlatan and forced[79] – all carry a deep political reading.

In 1920, just as the popularity of hunger artists waned, the *Washington Herald* published an article on a 'new world record' of resistance to inanition by Irish hunger striker and politician, Terence MacSwiney (1879–1920), in Brixton prison, London, which finally led to his death. The article mentioned: 'Dr Tanner, New York, 40 days, 1880; Griscom, Chicago, 31 days, 1881;[80] Signor Succi, New York, 45 days, 1890; Alexander Jacques, London, 50 days, 1891; Signor Merlatti, Paris, 50 days, 1886; Auguste Christensen, 35 days, 1901. The records bear other instances of fasting, but they are not authenticated. Several Irish prisoners in Cork jail claim to have fasted several days longer than MacSwiney.'[81] Inaccuracies apart, the coexistence of hunger artists and hunger strikers in the same league table is another sign of the inevitably political nature of fasting.

Perhaps hunger artists' performances seem milder when compared to the political strength of hunger strikes as powerful tools to fight for specific social aims. The force-feeding of women suffragettes or IRA members in the early decades of the twentieth century surely added more political charge to the act of fasting. Nevertheless, in the same way that any act of communication carries a strong political charge, the subtle, theatrical performances of our fasters also constituted a deep political phenomenon and an intrinsic part of the genuine *land of the hunger artists*.

Conclusion

Dr Robert Alexander Gunn, professor of surgery, dean of the US Medical College, and privileged witness of Henry Tanner's fast in New York, wrote his own 'deductions', the title of chapter VI of his *40 Days without Food*.[1] Even running the risk of anachronism, Gunn's deductions constitute, in my view, a tremendous source of inspiration for summarising the fundamental questions raised in this book. Of course, Gunn's conclusions on Tanner's fast should be placed in the context of 1880, and are neither comprehensive nor able to encompass all aspects of *the land of the hunger artists*. Despite this, they are enlightening for many of the issues that appear in the previous chapters and can help us, I think, to discuss the nature of the fasters' historical agency.

In Gunn's view, hunger artists could be a useful object of medical research and '[open] the door for a vast amount of study in the departments of *physiology, psychology, and hygiene*'.[2] Referring to Tanner's case (Figure 7.1), Gunn also pointed out the fact that fasting would reinforce experimentation on humans and self-experimentation as reliable practices that contributed towards overcoming contemporary reluctance about experiments with animals. Gunn therefore presented Tanner's fast as a pioneering event that could act as a mediator between physical, psychological, and hygienic cultures and practices, and fight against the rigidity of the medical profession, as he perceived the American medical class at that time. He also mentioned that hunger artists could challenge doctors' assumptions about the impossibility of fasting for forty days and therefore revisit medical consensus on the limits of human inanition.[3] He even dared to assert that 'the medical profession [had] yet much to learn, and that arrogant assumption which is so common in their ranks, should give place to a spirit of tolerant investigation of every new subject that presents itself for their consideration'.[4]

In Gunn's view, Tanner's fast – and we can take Tanner's case as a paradigm, as a source of inspiration for other later cases – could be a useful device to fight against the 'leading men of the profession',[5] a good example to fight a 'conservative' physiology. In Gunn's words:

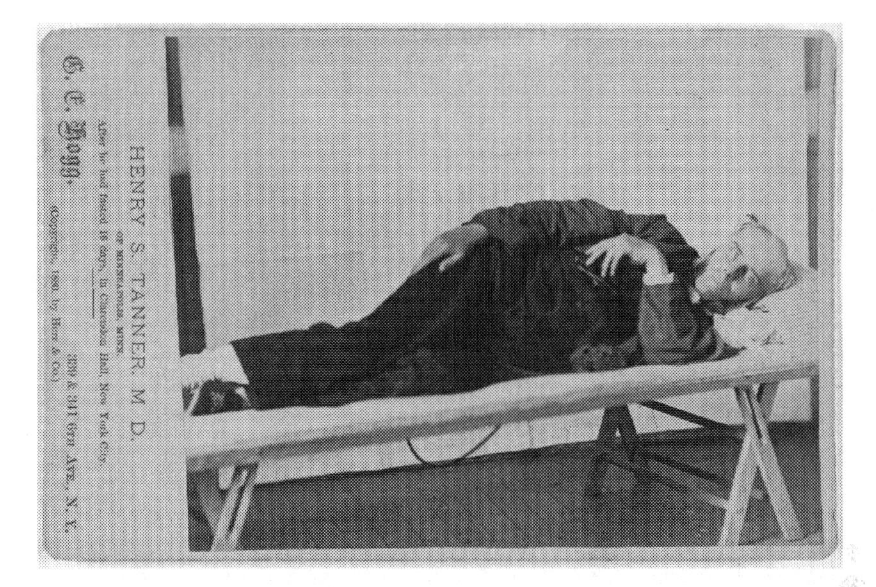

Figure 7.1. Portrait of Henry S. Tanner after fasting for sixteen days in Clarendon Hall, New York, 1880.

At first nearly all the prominent physicians throughout the country declared that a man could not fast forty days; that reported cases of such fasting [were] not authenticated ... Then when it became apparent that Dr Tanner was likely to succeed, they said it proved nothing, that many such cases were on record, and they had been of no benefit to science ... [Tanner] did not accept the teachings of the books, nor the authority of the so-called leading men in the profession. On the contrary, he had experimented for himself, and was certain, from absolute knowledge, what he could do. So, instead of the unknown 'irregular' doctor from Minneapolis learning from the medical scientists of the world, he has demonstrated that they are all in error, and must begin to study anew.[6]

In reality, controversy, the struggle for scientific authority and suspicion of fraud could be considered as obstacles to the functioning of the medical profession, or, on the contrary, as driving forces for further research and experimentation. A context of heterodoxy, in which hunger artists fitted well, could be perceived by historical actors themselves, but also from our historical perspective today, as a creative, stimulating factor. Gunn placed Tanner's fast in the context of an eclectic, popular medicine that gained new spaces of influence, at least in the United States, in the last decades of the century. It was precisely the commotion of the public sphere, the never-ending suspicion of deceit, the frequent disagreements between different schools of physiology, and the increase in controversial practices of the time, from homeopathy to spiritism, that left room for hunger artists' success as secular commodities, as another spectacle in urban fairs and freak shows, which benefited from the weaknesses of the established academic science. Moreover, fasting could prove, in Gunn's words, 'the wonderful power of mind over matter'[7] and pointed to the need for a more careful psychological study of these facts, at least in Tanner's view; he was convinced of the immortality of the soul and – like Succi and others – he was close to the spiritualist practices of the time.

Gunn also noticed that Tanner's public fast was an example of: 'the power of man to endure long fasts' and showed that 'persons who [were] ship-wrecked or otherwise deprived of food, would live longer than they do at the present time'.[8] In tune with future debates and the economy and the politics of hunger and fasting, he raised the issue that: 'as a rule *man eats too much*. It is an unquestionable fact that the custom of eating three times a day, gives the various organs of the body too much work to perform',[9] and insisted that: 'If we could learn a lesson in this particular, much force that is now spent in getting rid of the excess of the effete material of the body, might be saved and utilized for mental labor.' And he concluded that: 'It is a well known fact, that large eaters seldom have superior intellects.'[10] That statement opened the door again to a political reading of the minimum food intake, especially for the working classes, but also for the healthy citizens of the modern nation. As already discussed, hunger artists' spectacular performances and laboratory experiments were always entangled with issues of social justice, class struggle, cultural hegemony, and national pride. The very common assumption that people 'ate too much' had important consequences in industrial societies of the later nineteenth century and hunger artists acted as symbols and metaphors of the social tensions of the time that could be viewed from different perspectives as Caliban had masterfully described in *Le Figaro* in 1886.

Gunn concluded that in the near future more research and scientific papers would follow Tanner's successful self-experiment and fasting would progressively gain scientific authority: 'the lessons he [Tanner] will yet teach the world, as well as his memory, will live long after the chief of "regular" charlatanism shall have been forgotten'.[11] In fact, there is vast evidence that Tanner's fast in New York in 1880 resonated worldwide in the decades to come and became a sort of paradigm for many other hunger artists to discuss and imitate, as Succi's parade across the Brooklyn Bridge during his fast in 1890 clearly epitomised. What in 1880 was considered to be a foolish, charlatan-like attempt to live without food soon became a pattern to be followed and discussed in many European cities. From the American 'periphery' – in medical terms and referring to the contemporary prestigious schools of physiology in the old continent – Tanner's case conquered the global publics in the last decades of the nineteenth century in daily press articles, theatre plays, scientific papers, and popular leaflets.

Perhaps the most interesting point to discuss here is the fact that what concerned Gunn in 1880 is probably not too far from our present concerns about the historical value of hunger artists almost a century and a half later. Gunn's conclusions of Tanner's performance were probably biased by his personal involvement in the fasts and his professional sympathy for Tanner's eclectic medicine and values. Similarly, one might be tempted as a historian to get rid of the uncomfortable cacophony of all contemporary witnesses who spread multiple, often contradictory messages about those peculiar events. Things seem, however, more nuanced, subtle, and complicated. Gunn placed fasters and their performances as objects of scientific study between contemporary experimental physiology – with all its sophisticated instruments and quantification methods – the psychological turn – often linked to the anti-materialistic conceptions of the human body – and the public health, hygiene, and political dimension that issues such as hunger, fasting, inanition, and starvation had at that time. To that cocktail, one must add issues of hype, spectacle, commodity, and public commotion, which obviously hinder historical research, but provide a fluid landscape for the circulation of knowledge on fasting, inanition, and starvation that substantially enriches the overall picture.

Tanner stressed from the beginning of his career that fasting could have a powerful therapeutic effect, as an effective treatment for many illnesses of the digestive organs. In 1880, for example, the *British Medical Journal* concluded that Tanner's fast could be 'put to some use ... in case of severe dyspepsia, of gastric ulcer, and in the surgery of the abdominal cavity and alimentary canal starvation would seem to be a *perfectly rational means of treatment*, while it is worthy of trial in obesity'.[12] But the epistemological value of hunger artists actually transcended their strict therapeutic potential for digestive anomalies. As discussed in the previous chapters, hunger artists contributed to decentring human–animal research in many ways, especially in subjects related to fasting and inanition. Their performances fitted very well in the eclectic medical culture of the late nineteenth century, but at the same time they contributed to reinforcing it in controversial practices such as spiritism, homeopathy, hydropathy, and vegetarianism. Hunger artists and their controversial performances also shaped the way in which several medical disciplines, such as nutrition, evolved at the end of the century, and influenced the status of physiology as a supposedly unified discipline. As occurred with

Darwinism, thermodynamics, and other contemporary fields, fasts were therefore appropriated by different perspectives and professional interests. Doctors such as Luciani used Succi's case to try to keep physiology as a homogeneous discipline that could include philosophical and psychological factors. For Benedict and others, the study of hunger artists under conditions of prolonged inanition contributed to the robustness of the new science of nutrition and the studies of metabolism.

Equally, the fluid terrain of public commotion, controversy, and suspicion of fraud that surrounded hunger artists anywhere they travelled and performed seemed *a priori* a source of confusion and apparent irrationality. Nevertheless, it acted as a driving force for further medical research, protocol standardisation, and systematic control, which eventually led to new research and experimentation on processes of fasting, inanition, and starvation. In addition, the public commotion helps historians to take into account new, unexpected actors – the multiple and varied audiences of the hunger artists – who also had their say in circuits of knowledge in transit and in the continuous negotiations about authority, prestige, and credibility associated with them.

At the end, as a late nineteenth-century treatise on medical anomalies and curiosities pointed out, those experiencing and observing fasting processes always had something to gain in terms of contemporary knowledge about the functioning of the human body:

> The length of time which a person can live with complete abstinence from food is quite variable. Hippocrates admits the possibility of fasting more than six days without a fatal issue; but Pliny and others allow a much longer time, and both the ancient and the modern literature of medicine are replete with examples of abstinence to almost incredible lengths of time … In many religions fasting has become a part of worship or religious ceremony … Exhibitionists … either for notoriety or for wages, demonstrate their ability to forgo eating, and sometimes drinking, for long periods. Some have been clever frauds, who by means of artifices have carried on skilful deceptions; others have been really *interesting physiologic* anomalies.[13]

Topography

Through the overlapping of the different chapters, I have provided a sort of travel guide for 'visitors' to the *land of the hunger artists*, that uncertain and often uncharted territory. This travel guide shows convincingly, I hope, that those who followed their rules and procedures – although controversial – gained public recognition and authority. Long itineraries through cities, regions, nations, and continents, and the numerous local contingencies associated with the culture of travel provide a global character to hunger artists' controversial careers. What in some places gained the applause of the established medical community in others became bitter criticism of the charlatan nature of the fasters and the supposed lack of interest they held for scientific study. This was therefore an intrinsic part of the game. In 1890, for example, when Succi was enjoying great success in his fast in New York, the *British Medical Journal* stated that 'The faster, the acrobat, the professional Hercules, the parachutist – are all in the same category; they are people who by training or natural endowment are able to accomplish what is beyond the power of ordinary men and women.'[14] It was therefore that 'heroic' nature of hunger artists, able to explore some of the limits of the human body, that contributed to making sense of a peculiar, controversial, apparently banal profession.

The topography of the land included colonial, exotic adventures, such as Succi's trips to Africa, and a rich variety of urban contexts, which transcended the big cities – Paris, London, New York – and progressively extended to secondary cities in Europe and the Americas. And, once in the city, the general map became a more detailed urban plan that often associated the great monuments and civic sites with the presence of hunger artists in the science pavilions of international exhibitions, towers, cathedrals, theatres, cinemas, parks, avenues, the Eiffel Tower, Brooklyn Bridge, and other emblematic places that acted as active containers of the performances and shaped the contemporary knowledge of hunger, inanition, and starvation in different directions.

In addition, the 'laws' of the land of the hunger artists were always under permanent discussion, from the contracts signed between hunger artists and their impresarios, to local regulations for public spectacles and the 'laws' of experimental physiology in laboratories and medical academies, which ranged from suspicion of fraud, willingness of medical supervision, and dismissal of hunger artists based on their lack of scientific interest. There were, though, tacit agreements on the rules of the game that transcended local contingencies and cultures: sealed cages, contracts, medical boards, overnight supervision, the implicit 'control' of the daily press, celebrations at the end of the fasts, crowded public lectures, advertisements, patent medicines, controversial elixirs, and mineral waters. This whole package travelled across the land, from bourgeois elites to the lowest classes, as a provocative commodity.

In our imaginary topography, actors, places, objects, and rituals acted as mountains, rivers, and valleys of the land of the hunger artists. They comprised an intricate network that resulted in a peculiar culture of fasting from the 1880s to the 1920s. Celebrities such as Tanner, Succi, Merlatti, Levanzin, and others, plus their audiences – doctors such as

Luciani, Maréchal, Richet, Benedict, Rubner, and Morgulis; science popularisers such as Figuier, Meunier, Fonvielle, and Parville; journalists such as Rahola and Saavedra and many anonymous authors of articles in the daily press; impresarios such as Francis Train, Lamperti, and Ducazcal; artists, medical students, visitors of all sorts (including many women) – gained public authority and recognition through their travels, their synchronic appearances in the daily press worldwide, their spectacular and controversial performances, their public displays in unexpected places, their capacity to stir public commotion, their submission to scientific experiments and the sophisticated instruments of 'modern' physiology, their ingestion of elixirs, patent medicines, and mineral waters with supposedly miraculous properties. All these actors, practices and places constituted a good part of the identity of the *land of the hunger artists*.

This land also had a very rich material culture: sealed cages and laboratory calorimeters became two habitual sites for hunger artists' confinement, experimentation, and spectacle. Other physiology instruments attempted – not always successfully – to quantify and mathematise the process of inanition. Fences and windows separated hunger artists from their visitors but at the same time enabled conversation between them. Elixirs, homeopathic drugs, and mineral waters – in particular the tremendous controversy around Succi's liquor – brought hunger artists close to traditional charlatan practices, but gave them a spectacular public dimension. Swords for fencing spectacles, advertisement posters, and signed photographs fitted very well in a framework of commodities to be bought and sold in the urban marketplaces of the time, and reinforced the secular character of the hunger artists over the old mystical and religious traditions of abstinence and asceticism. Supervisory medical boards' reports, blood and urine samples, and the doctors' systematic supervision acted as a powerful tool against potential fraud and reinforced the scientific authority of the medical class before a public commotion that often became public confusion.

Spiritist sessions, hypnotic dreams, fictional narratives, periods spent in mental asylums, gendered anorexia nervosa, and ambiguous explanations of vital forces and willpower were all closely associated with hunger artists' fasts; they opened the door for non-materialistic explanations of the causes of human resistance to inanition and contributed to the psychological turn of the last decades of the century. They proved, as discussed earlier, that it was the heterodox character of hunger artists that in practice constructed their social authority and prestige, their relevant impact, and contemporary scientific debates. Luciani's attempt, among those of many others, to keep physiology as a unified, holistic discipline inevitably opened the door to new 'psychological' explanations of hunger, inanition, and starvation. Hunger artists appropriated the fissures of anti-materialism to advance their own professional interest.

As those late nineteenth-century witnesses of hunger artists' performances – including Kafka himself – hardly achieve consensus on issues such as experimentation, fraud, objectivity, materialism, hunger, inanition, and willpower – just to mention some of the key concepts that were under permanent discussion – our present historical approach is obviously also far from any standardised 'objectivity'. Nevertheless, the blurred boundaries of the fasts of Tanner, Succi, and many others provide a tremendous explanatory power of the complexities of the past. It is precisely this rich topography that helps us to describe the subtleties and contradictions. As a whole, in the seamless webs of the land of the hunger artists, travels, performances, experiments, heterodox practices, public commotion, and exhibition complexes endorsed their continuous attempt to gain authority, recognition, and prestige. It was precisely in that peculiar public sphere that our professional fasters acquired the status of epistemologically relevant historical actors.

Decline

In 1920, just two years before the appearance of Kafka's short story, the Argentinian doctor Luis C. Maglioni published a leaflet on his self-experimentation in a 37-day fast, after having acquired great experience in hospitals in different European cities.[15] Maglioni can be considered as a late 'citizen' of the land of the hunger artists, with his multi-cultural encounters, his itinerant journeys, and his peculiar role as self-experimenter – à la Tanner – being both faster and witness. Maglioni had worked as a surgeon in Argentina and later in several hospitals in Paris, Zurich, Berlin, and London. He had a broad range of interests, including surgery, hypodermic therapeutics, fasting, hunger strikes, politics, and art.[16]

As a privileged witness and protagonist, Maglioni provided an in-depth description of his own self-experiments, his self-fasts, first in Berlin in 1911, and later in Paris in 1914 and London in 1915. He dedicated his leaflet to Edward H. Dewey, whose works on therapeutic fasting had guided him throughout his career.[17] Maglioni also mentioned some of the most famous 'speculative, lucrative' fasts – Succi, Merlatti, and more. In his leaflet, he first described the main fasts he had witnessed around 1900. Then, he addressed the resistance to inanition by sick persons, and the therapeutic uses of fasting. Finally, he commented on 'political fasters', since he dedicated the booklet to Terence MacSwiney, and to eleven Irish hunger strikers in the prisons of Brixton and Cork. It was also in 1920, during the Irish War of Independence, that MacSwiney, at that time an elected Sinn Féin mayor of the city of Cork who had been arrested by the British government and imprisoned in Brixton prison, began a hunger strike as a protest against the charges of sedition.[18] The campaign, as in the hunger artists' case, but now with additional political dimensions, soon acquired a notable international resonance. After several attempts at force-feeding,[19] which he strongly refused, MacSwiney died from starvation in the prison in October that year. His case became an example for future Irish hunger strikes, in particular those of 1923.

In 1920, when Maglioni wrote his leaflet, *Mis 37 días de ayuno (auto-experiencia)*, the popularity of hunger artists' performances had notably declined. Maglioni's recapitulation of former fasting events and experiences and his deep interest in new political dimensions of fasting were probably a sign of the decline of the golden years. Indeed, in the early decades of the twentieth century, as happened with the Irish pro-independence movement and also with the women suffragettes,[20] hunger strikes acted as powerful political tools to bring shame down on the oppressor and capture the sympathy and solidarity of co-religionists. Terrible images of force-feeding in prisons also contributed to this turn towards a new social and political dimension of the old art of fasting.[21] In addition, perhaps not by chance, just two years later, Kafka published his short story, in which the main character, a hunger artist, was expelled from his cage amidst the growing indifference of the audience. The audience now asked for other kinds of excitement and entertainment: the black panther that replaced the artist in the cage, but also the new wonders of electricity, cinematography, automobiles, early aeroplanes, and all the technological fascination of a new modernity that had progressively transformed the old freak show culture and the Benjaminian 'arcades' into truly historical relics.[22]

Moreover, the mortality rates, famines, and enormous human suffering provoked by the First World War and, afterwards, by the terrible consequences of the Spanish flu epidemic changed the scale of values and world views of a Western population that now perceived entertainment and hype associated with hunger and fasting as a joke in bad taste.[23] In a way, what the journalist Federico Rahola had already denounced in Barcelona in 1888 seems to come crudely true in 1920. When famine, real hunger, and starvation become a plague for a nation, there is little room for the frivolous exhibition of emaciated bodies in inanition in a theatrical manner.[24]

But the decline of hunger artists also went hand in hand with the decline of the therapeutic uses of fasting. In 1923, just one year after Kakfa's story and five years after the death of Tanner and Succi, the Nobel Prize in Physiology or Medicine was awarded jointly to Frederick Grant Banting and John James Rickard Macleod for the discovery of insulin. Until that date, fasting had kept the original therapeutic aims that Tanner had already proclaimed in 1880, just at the beginning of our period of study. But, in the same way that Kakfa's hunger artist was abandoned by the audience in his cage, the therapeutic use of fasting to fight diabetes, childhood diseases (rickets), epilepsy, and even obesity also began to fade. From the 1920s onwards, therapeutic fasting was progressively considered to be a risky, unsafe method for patients and was therefore abandoned.[25]

Much research is needed to refine the former arguments and to better clarify the reasons for the decline of hunger artists. Accepting the never-ending tensions between continuity and change in history, one could even show some evidence of a certain renaissance of fasting performances in recent times and the reappearance of fasting practices and diets for the improvement of human health and wellbeing.[26] Nevertheless, the time in which thousands (and probably millions on a global scale) witnessed or heard about the heroic performances of a small group of men (and women) who exhibited their bodies in public places and private laboratories for their own profit, as well as for the curiosity of many and the scientific interest of few, will never come back. The land of the hunger artists of the late nineteenth century has irreversibly disappeared from the cartography of our present, but remains as a powerful historical lesson that we should add to our knowledge of the past. It teaches us about how that spectacle raised questions about objectivity, the constraints of experimentation, the validity of materialism, the nature of scientific authority, and the boundaries between science and entertainment.

Perhaps hunger artists could be deleted from the history of hunger, from the historical investigation of experimental physiology, or even from the standard accounts of hype and spectacle in urban, industrial societies of the late nineteenth century. However, in doing so, we would lose genuine, relevant, irreplaceable world views. Hunger artists are useful, necessary bridges to understand a considerable part of the Western culture of the fin-de-siècle: from growing scientific optimism on a global scale to the spread of heterodox practices on a local level; from urban reforms and the materiality of a rhetoric of modernity to the public success of the freak culture and hype that protected hunger artists in their times of glory; from the pleasures of the bourgeois gastronomy to the austerity and disciplinary challenges that hunger and inanition posed for contemporary witnesses of public fasts.

There is no doubt that *the land of the hunger artists* provides an incomparable documentary richness which has resulted in this book, but still deserves future historical analysis.

Notes

Preface and Acknowledgements

1 Pere Hereu, et al., *Arquitectura i ciutat a l'Exposició Universal de Barcelona de 1888* (Barcelona: Universitat Politècnica de Catalunya, 1988).
2 Federico Rahola, 'Carta a Succi', *La Vanguardia*, 15 Oct. 1888, 2.
3 Agustí Nieto-Galan, 'Scientific "Marvels" in the Public Sphere: Barcelona and Its 1888 International Exhibition', *HoST – Journal of History of Science and Technology*, 6 (2012), 33–63.
4 Breon Mitchell, 'Kafka and the Hunger Artists', in Alan Udoff (ed.), *Kafka and the Contemporary Critical Performance: Centenary Readings* (Bloomington: Indiana University Press, 1987), 238.
5 Franz Kafka, *A Hunger Artist and Other Stories: Translated by Joyce Crick, with an Introduction and Notes by Ritchie Roberts* (Oxford: Oxford University Press, 2012 [1922]); Mitchell, 'Kafka and the Hunger Artists'.
6 Kafka relied on his own observations, newspapers, and contemporary scientific studies to write the short story: Mitchell, 'Kafka and the Hunger Artists', 238.
7 Succi's birthdate is uncertain; according to different sources, it ranges from 1851 to 1853.
8 Joan Jacobs Brumberg, *Fasting Girls: The Emergence of Anorexia Nervosa as a Modern Disease* (Cambridge, Mass.: Harvard University Press, 1988).
9 Maud Ellmann, *The Hunger Artists: Starving, Writing and Imprisonment* (London: Virago Press, 1993).
10 Robert Bogdan, *Freak Show: Presenting Human Oddities for Amusement and Profit* (Chicago: University of Chicago Press, 1988). See also Agustí Nieto-Galan, 'Hunger Artists and Experimental Physiology in the Late Nineteenth Century: Mr Giovanni Succi Meets Dr Luigi Luciani in Florence', *Social History of Medicine*, 28 (1) (2015), 82–107.
11 Nieto-Galan, 'Hunger Artists and Experimental Physiology'.
12 Oliver Hochadel and Agustí Nieto-Galan (eds.), *Barcelona: An Urban History of Science and Modernity (1888–1929)* (London: Routledge, 2016); Oliver Hochadel and Agustí Nieto-Galan (eds.), *Urban Histories of Science: Making Knowledge in the City* (London: Routledge, 2019).
13 Agustí Nieto-Galan, 'Useful Charlatans: The Fasting Contest of Giovanni Succi and Stefano Merlatti in Paris, 1886', *Science in Context,* 33 (4) (2020), 405–422.
14 Historical Epistemology Seminar, Universidad Nacional de Educación a Distancia, Madrid (2014); BSHS Conference, St Andrews (2014); HSS Annual Meeting, Chicago (2014); the international workshop 'Les mises en

scène des sciences et leurs enjeux, 19e–21e siècle' at the Institut Historique Allemand, Paris (2015); the conference 'The Pursuit of Global Urban History', Leicester, UK (2019). I have also presented a draft version of this book at the Global History of Science Seminar (GHOSS), in my Institut d'Història de la Ciència (iHC) in Barcelona. More recently, I have given two lectures on hunger artists in Milwaukee (Department of Languages, Literatures, and Cultures, Marquette University) and Madison (Latin American, Caribbean and Iberian Studies (LACIS) Program, and The Robert F. and Jean E. Holtz Center for Science and Technology Studies).

15 Domenico Priori, 'Il digiunanote e lo scienziato', *Rendiconti Accademia Nazionale delle Scienze detta dei XL. Memorie di Scienze Fisiche e Naturali*, 135 (2017), 1–9.

Introduction

1 Ernest Monin and Philippe Maréchal, *Stefano Merlatti. Histoire d'un jeûne célèbre, précédée d'une étude anecdotique, physiologique et médicale sur le jeûne et les jeûneurs* (Paris: C. Marpon et E. Flammarion, 1887).

2 Louis Figuier, 'Les nouveaux jeûneurs: Succi et Merlatti', *L'Année scientifique et industrielle*, 30 (1886), 395–404.

3 'Oh! Malheureuse Italie! Tu nous avais donné la succulente mortadelle et le doux macaroni, et tu nous envois maintenant l'amertume avec Succi et Merlatti! ... Si les fatales doctrines de des jeûneurs se propageaient, messieurs, c'en serait fait de l'art culinaire, le principe de toutes les choses, le père de tous les autres arts. Ce serait fini, l'homme étouffant toutes ses aspirations vers le bon et le meilleur, ne serait plus l'homme, il ne serait plus même un végétal, car les végétaux se nourrissent, mais un simple minéral! ... Luttons de toutes les forces de nos estomacs, organisons partout des comités de résistance? Jurons de ne jamais nous quitter avant complète indigestion': Albert Robida, 'La jeûnomanie', *La Caricature*, 363 (1886), 406; quoted in Agustí Nieto-Galan, 'Useful Charlatans: The Fasting Contest of Giovanni Succi and Stefano Merlatti in Paris, 1886', *Science in Context*, 33 (4) (2020), 405.

4 Francisque Sarcey (1827–1899) was a French journalist and drama critic, fond of vegetarianism and defender of the therapeutic virtues of diet and fasting.

5 Victorien Sardou (1831–1908) was a dramatist. He wrote the play *Fedora* (1882) expressly for Sarah Bernhardt. Félix Duquesnel (1832–1915) was a playwright, novelist, and theatre manager.

6 Pierre Baour-Lormian (1770–1854) was a poet and writer.

7 Hunyadi Janos was a mineral water Succi drank during his fast (see Chapter 5).

8 'En présence du succès obtenu par Succi et Merlatti, Sarah Bernhardt et Coquelin, qui ne reculent devant aucun obstacle, se sont engagés à jeûneur chacun trente jours. Ce n'est pas un jeûne com les autres. Au lieu de se priver de nourriture, ces deux grands artistes se privent de réclames. Ils promettent de demeurer un mois sans faire parler d'eux dans aucun journal. Il y a déjà huit jours que Coquelin demeure immobile sous la surveillance implacable

du sévère Sarcey. Quant à Sarah, elle reste concentrée et silencieuse entre les mains de M. Sardou et M. Duquesnel. Le neuvième jour, Sarah a commencé à frétiller ... On redoute une réclame pour demain dans l'*Evènement.* Coquelin tient bon. Il est vrai qu'on lui a promis qu'il serait décoré le 14 juillet prochain. Sarah Bernhardt aussi. *Bulletin de Coquelin:* Pouls: 114 pulsations; Dynamomètre: a résisté à une tragédie de Baour Lormian. Haleine: agréable. A bu un verre d'eau qu'il a rendu instantanément. *Bulletin de Sarah Bernhardt:* Pouls: 3 pulsations; Dynamomètre: Ecrasement de Mlle Noirmont d'un seul coup de poing; Haleine: parfumé; Poids: 0,5 centigrammes. A bu un verre d'eau Hunyadi Janos, qu'elle a rendu involontairement': 'Le jeune de Coquelin ainé et de Sarah Bernhardt', *Le Figaro*, 1 Jan. 1887, 3.

9 Angiolo Filippi (1836–1905) studied medicine in Pisa and worked as a prison doctor before joining Garibaldi's red shirts in the campaign for the independence and unification of Italy. In 1884 he attained a chair of legal medicine in Florence and, in 1889, published a widely read textbook on forensic medicine. He founded a laboratory for medico-legal research in Florence. See *Nature*, 26 Sep. 1936, 538.

10 Angiolo Filippi, 'Il Sor Giovanni Succi. Digiunatore e l'Academia Medico Fisica Fiorentina', *Lo Sperimentale*, 61 (1888), 325.

11 Walter Vandereycken and Ron van Deth, *From Fasting Saints to Anorexic Girls: The History of Self-Starvation* (London: Athlone Press, 1994); Peter Payer, *Hungerkünstler eine verschwundene Attraktion* (Vienna: Sonderzahl, 2000); Peter Payer, *Hungerkünstler in Wien. Eine verschwundene Attraktion* (Vienna: Sonderzahl, 2002); Sharman Apt Russell, *Hunger: An Unnatural History* (New York: Basic Books, 2005); Todd Tucker, *The Great Starvation Experiment: The Heroic Men Who Starved So That Millions Could Live* (New York: Free Press, 2006); James Vernon, *Hunger: A Modern History* (Cambridge, Mass.: Belknap Press, 2007); Ian Miller, *A Modern History of the Stomach: Gastric Illness, Medicine and British Society, 1800–1950* (London: Pickering & Chatto, 2011).

12 Elizabeth Neswald and David F. Smith. *Setting Nutritional Standards: Theory, Policies, Practices* (Rochester, N.Y.: Studies in Medical History, 2017).

13 Erika Dyck and Larry Stewart, *The Uses of Humans in Experiment: Perspectives from the Seventeenth to the Twentieth Century* (Amsterdam: Clio Medica, 2016).

14 Elizabeth Williams, *Appetite and Its Discontents: Science, Medicine and the Urge to Eat, 1750–1950* (Chicago: University of Chicago Press, 2020).

15 Emma C. Spary, *Eating the Enlightenment: Food and the Sciences in Paris* (Chicago: University of Chicago Press, 2012); Emma Spary and Anya Zilberstein (eds.), 'Food Matters', *Osiris*, 35 (2020). See also Maud Ellmann, *The Hunger Artists: Starving, Writing and Imprisonment* (London: Virago Press, 1993); Nina Diezemann, *Die Kunst des Hungerns. Essstörungen in Literatur und Medizin um 1900* (Berlin: Kadmos, 2006).

16 Payer, *Hungerkünstler in Wien.* Agustí Nieto-Galan, 'Scientific "Marvels" in the Public Sphere: Barcelona and Its 1888 International Exhibition', *Host – Journal of History of Science and Technology*, 6 (2012), 7–38.

17 Vandereycken and van Deth, *From Fasting Saints to Anorexic Girls*; Ernest-Charles Lasègue, 'De l'anorexie hystérique', *Archives générales de médicine*, 1873, 316–386; William Withey Gull, 'Anorexia Nervosa', *Clinical Society's Transactions*, 7 (1874), 22.

18 Steven Shapin, '"You Are What You Eat": Historical Changes in Ideas about Food and Identity', *Historical Research*, 87 (2014), 377–392. As Maud Ellmann wrote some years ago: 'hunger exemplifies the fact that the body is determined by its culture, because the meanings of starvation differ so profoundly according to the social context in which it is endured ... hunger can be caused by anything from famine, war, revolution, disease, psychosis, dieting or piety' (Ellmann, *The Hunger Artists*, 4).
19 Aileen Fyfe and Bernard Lightman (eds.), *Science in the Marketplace: Nineteenth-Century Sites and Experiences* (Chicago: University of Chicago Press, 2007).
20 See, for example, Jean-Louis Flandrin and Massimo Montanari (eds.), *Food: A Culinary History from Antiquity to the Present* (New York: Columbia University Press, 1999); Carole Counihan and Penny van Esterik (eds.), *Food and Culture: A Reader* (New York: Routledge, 1997); Ursula Klein and Emma C. Spary, *Materials and Expertise in Early Modern Europe: Between Market and Laboratory* (Chicago: University of Chicago Press, 2009).
21 'Food: The Forgotten Medicine'. This title was used for a June 2016 conference at the Royal Society of Medicine: www.collegeofmedicine.org.uk/events/#!event/2016/6/8/food-8211-the-forgotten-medicine (accessed 2 Jun. 2016). See also Susan E. Cayleff, *Nature's Path: A History of Naturopathic Healing in America* (Baltimore: Johns Hopkins University Press, 2016).
22 'Spaces of food procurement, preparation and consumption are also spaces of knowledge production and circulation': Spary and Zilberstein (eds.), 'Food Matters', 15.
23 Corinna Treitel, *Eating Nature in Modern Germany: Food, Agriculture and Environment, c. 1870 to 2000* (Cambridge: Cambridge University Press, 2017).
24 Charles Richet, *Physiologie. Travaux du laboratoire de M. Charles Richet. Chimie physiologique. Toxicologie* (Paris: Félix Alcan, 1893–1895), vol. II, 301.
25 Norton Wise, 'Mediating Machines', *Science in Context*, 2 (1) (2008), 77–113.
26 Thomas Gieryn's diagnosis of the processes of demarcation of scientific knowledge and authority can be applied to the analysis of hunger artists' performances and the attribution of a 'scientific' status to that kind of fact. See Thomas F. Gieryn, 'Boundary-Work and the Demarcation of Science from Non-Science: Strains and Interests in Professional Ideologies of Scientists', *American Sociological Review*, 48 (6) (1983), 781; Thomas Gieryn, *Cultural Boundaries of Science: Credibility on the Line* (Chicago: University of Chicago Press, 1999).
27 '[P]rotester contre la science des jeûneurs et des jeûneuses qui devient excessivement menaçante en ce moment': Wilfrid de Fonvielle, *Mort de faim. Étude sur les nouveaux jeûneurs* (Paris: Librairie illustrée, 1886), 3.
28 '[L]a science médicale française est bien malade': ibid., 63.
29 Robert Darnton, *Mesmerism and the End of the Enlightenment in France* (Cambridge, Mass.: Harvard University Press, 1968).
30 Robert Fox, *The Savant and the State: Science and Cultural Politics in Nineteenth-Century France* (Baltimore: Johns Hopkins University Press, 2012).
31 Fonvielle, *Mort de faim*, 8.
32 'En ayant recours aux lumières, que la science met déjà à notre disposition, nous pensons qu'il est possible de se faire une idée exacte des phénomènes qui accompagnent la faim, sans avoir recours aux comités de surveillance, ...

sans nous laisser influencer le moins du monde par le résultat des expériences que nous n'accepterons que si elles affirment des choses qui soient conformes avec ce que l'ensemble de la physiologie nous a appris. En effet, en procédant autrement, notre science serait toujours précaire et sujette à donner des résultats peu dignes de la peine qu'on se donne pour les assurer': ibid.

33 Sergius Morgulis, *Fasting and Undernutrition: A Biological and Sociological Study of Inanition* (New York: Dutton and Company, 1923), 9.

34 Richard Altick, *The Shows of London* (Cambridge, Mass.: Belknap Press, 1978); Iwan Rhys Morus, *Frankenstein's Children: Electricity, Exhibition, and Experiment in Early Nineteenth-Century London* (Princeton, N.J.: Princeton University Press, 1998).

35 Robert Bogdan, *Freak Show: Presenting Human Oddities for Amusement and Profit* (Chicago: University of Chicago Press, 1988).

36 Sigal Gooldin, 'Fasting Women, Living Skeletons and Hunger Artists: Spectacles of Body and Miracles at the Turn of a Century', *Body and Society*, 9 (2) (2003), 39.

37 R. H. Park and M. P. Park, 'Goya's Living Skeleton', *British Medical Journal*, 304 (1992–3), 844.

38 National Fairground and Circus Archives. Studio carte de visite photograph of 'The Living Skeleton' George Prise, 178c81.46.

39 Altick, *The Shows of London.*

40 Firmin Maillard, *Recherches historiques sur la morgue* (Paris: A. Delahays, 1860).

41 Matthew Ramsey, 'Sous le régime de la législation de 1803. Trois enquêtes sur les charlatans au XIXe siècle', *Revue d'histoire moderne et contemporaine*, 27 (1980), 485–500; Roy Porter, *Quacks, Fakers and Charlatans in English Medicine* (Stroud: Tempus, 2000); Tal Golan, 'Blood Will Out: Distinguishing Humans from Animals and Scientists from Charlatans in the Nineteenth-Century American Courtroom', *Historical Studies in the Physical and Biological Sciences*, 31 (1) (2000), 93–124; Pope Brock, *Charlatan: The Fraudulent Life of John Brinkley* (London: Weidenfeld & Nicolson, 2008); Irina Podgorny (ed.), *Charlatanes. Crónicas de remedios incurables* (Buenos Aires: Eterna Cadencia Editora, 2012).

42 Irina Podgorny and Daniel Gethmann, '"Please, Come In": Being a Charlatan, or the Question of Trustworthy Knowledge', *Science in Context*, 33 (4) (2020), 355–361.

43 Michael Hagner, 'Scientific Medicine', in David Cahan (ed.), *From Natural Philosophy to the Sciences: Writing the History of Nineteenth-Century Science* (Chicago: University of Chicago Press, 2003), 49–87; Richard L. Kremer, 'Physiology', in Peter Bowler and John V. Pickstone (eds.), *The Cambridge History of Science*, vol. VI, *The Modern Biological and Earth Sciences* (Cambridge: Cambridge University Press, 2009), 342–366; John Pickstone, 'Physiology and Experimental Medicine', in Robert Olby, et al. (eds.), *Companion to the History of Science* (London: Routledge, 1990), 728–742.

44 Kremer, 'Physiology'.

45 Jean-Hervé Lignot and Yvon LeMaho, 'A History of Modern Research into Fasting, Starvation and Inanition', in Marshall D. McCue (ed.), *Comparative Physiology of Fasting, Starvation, and Food Limitation* (Dordrecht: Springer, 2012), 7–23.

46 John Warne Monroe, *Laboratories of Faith: Mesmerism, Spiritism and Occultism in Modern France* (Ithaca: Cornell University Press, 2008).
47 Hagner, 'Scientific Medicine', 86.
48 Martin Dinges, et al. (eds.), *Medical Practice, 1600–1900: Physicians and Their Patients* (Leiden: Brill, 2016).
49 Kremer, 'Physiology', 358.
50 '[I]n the newer histories, the spread of an independent experimental physiology has become much more complex than a simple transplantation of French or German models to universities in Britain, the US and elsewhere': ibid., 355.
51 'From the vantage point of a single part of the world or of *powerful elites*, but rather widens [its] scope, socially and geographically, and introduces *plural voices* into the account': Natalie Zemon Davis, 'Decentering History: Local Stories and Cultural Crossings in a Global World', *History and Theory*, 50 (2) (2011), 190 (my emphasis).
52 David Livingstone and Charles Withers, *Geographies of Nineteenth-Century Science* (Cambridge: Cambridge University Press, 2011).
53 David Livingstone, *Putting Science in Its Place: Geographies of Scientific Knowledge* (Chicago: University of Chicago Press, 2003); Livingstone and Withers, *Geographies of Nineteenth-Century Science.*
54 Sven Dierig, Jens Lachmund, and J. Andrew Mendelson (eds.), 'Science and the City', *Osiris*, 18 (2003).
55 John S. Haller, *Medical Protestants: The Eclectics in American Medicine, 1825–1939* (Carbondale: Southern Illinois University Press, 1994).
56 See, for example, Robert Jütte, 'The Historiography of Nonconventional Medicine in Germany: A Concise Overview', *Medical History*, 43 (1999), 342–358.
57 Monroe, *Laboratories of Faith*, 4–10. On scientific objectivity, see Lorraine Daston and Peter Galison, *Objectivity* (New York: Zone Books, 2007).
58 'Unable to disentangle them from their botanic biases and ill-equipped to accept the implication of germ theory, the financial costs of salaried faculty and staff, or the research implications of laboratory science, the eclectics emerged into the twentieth century stubborn and fundamentally perplexed by the rigor demanded in the new medical curriculum and the expectation of research': Haller, *Medical Protestants*, xix.
59 Robert Bivins and Hilary Marland, 'Weighing for Health: Management, Measurement and Self-Surveillance in the Modern Household', *Social History of Medicine*, 29 (4) (2016), 757–780.
60 As Anita Guerrini rightly pointed out, 'Human experimental subjects have only recently become subjects of historical scrutiny, as knowledge making has come to be viewed more broadly': Anita Guerrini, 'The Human Experimental Subject', in Bernard Lightman (ed.), *A Companion to the History of Science* (Chichester: Wiley-Blackwell, 2016), 136.
61 Laurence Altman, *Who Goes First? The Story of Self-Experiments in Medicine* (Berkeley: University of California Press, 1998).
62 Peter J. Atkins, Peter Lummel, and Derek J. Oddy (eds.), *Food and the City in Europe since 1800* (Aldershot and Burlington: Ashgate, 2007).
63 Anson Rabinach, *The Human Motor: Energy, Fatigue and the Origins of Modernity* (Berkeley: University of California Press, 1992), 85.

1 Geographies

1 Federico Minutilli, *La Tripolitania* (Turin: Fratelli Bocca, 1902).
2 *Proceedings of the Royal Geographical Society*, 4 (1882), 244. See also *L'Exploration. Revue des conquêtes de la civilisation sur tous les points du globe,* 13 (1882), 567–568, www.treccani.it/enciclopedia/federico-minutilli/ (accessed 10 Mar. 2019).
3 'In 1879, Signor Succi was in the vicinity of the Voami River at the same time that Stanley was exploring the district': *The Man of Mystery. The Fasting Man of the Nubian Desert* (1890), Bodleian Libraries, University of Oxford, John Johnson Collection: Entertainments Folder 9 (44).
4 Federico Minutilli, 'Nel mar delle Indie. Viaggio di un italiano', *Nuova Antologia*, 32 (1882), 108–113; 'Il viaggio d'esplorazione commerciale di Giovanni Succi', *Bolletino della Sociëta geogràfica italiana*, 16 (4) (1882), 350–354. Geographical societies often placed Succi among the great Italian names. In 1883, for example, the Ain Geography Society noted that: 'From Zanzibar to Suez, the main explorations were conducted by Italians. Already last year ... the Italian government [sent] ... a real multitude of explorers to the coasts of the Red Sea: Bianchi, Licata, Cecchi, Succi, Naretti ... [to] establish a trade route between inland and the Italian colony of Assab.' See *Bulletin de la Société de Géographie de l'Ain*, 1 (1883), 286. Assab is a port city in the Southern Red Sea region of Eritrea and a former Italian colony. See also Sofia Bompiani, *Italian Explorers in Africa* (London: Religious Tract Society, 1891).
5 Valeska Huber and Jürgen Osterhammel (eds.), *Global Publics and Their Limits, 1870–1990* (Oxford: Oxford University Press, 2020).
6 See for example Simon Winchester, *Krakatoa: The Day the World Exploded, 27 August 1883* (London: Viking, 2003).
7 Jürgen Osterhammel and Patrick Camiller, *The Transformation of the World: A Global History of the Nineteenth Century* (Princeton, N.J.: Princeton University Press, 2014).
8 'Where things happen is crucial to knowing how and why they happen': Barney Warf and Santa Arias (eds.), *The Spatial Turn: Interdisciplinary Perspectives* (London: Routledge, 2009), 1; David Livingstone, *Putting Science in Its Place: Geographies of Scientific Knowledge* (Chicago: University of Chicago Press, 2003).
9 Elizabeth Neswald, 'Strategies of International Community-Building in Early Twentieth-Century Metabolism Research: The Foreign Laboratory Visits of Francis Gano Benedict', *Historical Studies in the Natural Sciences*, 43 (1) (2013), 1–40.
10 For nineteenth-century travels to the Comoran islands, see Barbara Dubins, 'Nineteenth-Century Travel Literature on the Comoro Islands: A Bibliographical Essay', *African Studies Bulletin*, 12 (2) (1969), 138–146. See also Nikolaos Mavropoulos, 'The Japanese Expansionism in Asia and the Italian Expansion in Africa: A Comparative Study of the Early Italian and Japanese Colonialism', PhD, La Sapienza, Rome, 2019.
11 Giovanni Succi, *Commercio in Africa. Il Madagascar, l'isola di Johanna e l'arcipelago di Comoro, Zanzibar e Mozambese* (Milan: Tipografia Nazionale, 1881).

12 'Empty Bread Baskets as a Steady Diet', *New York Herald*, 9 Nov. 1890.
13 'Purchase of export goods from Africa: sugar, leather, elastic gum, carnations, mother of pearl, wax. Import: manufacturing of cotton, trinkets from Venice, flour from Trieste … Therefore, on the evidence of similar advantages based on real results of operations already carried out on my travels, I have no doubt that the great Italian Commerce will want to get an idea of my offering, and that I will see a respectable and energetic company to undertake such an enterprise; a great but risk-free enterprise, since nothing can raise the slightest doubt about the safety of success', Milan, 24 Nov. 1881: Succi, *Commercio in Africa*, 28. 'Jean Succi wishes to establish a Commercial house hereupon this island. I hereby agree to give them a free port for the landing and storing and shipping of coal, and all goods brought here in transit shall be free of all duties and taxes. Goods landed to be consumed upon the island shall pay a duty of five per cent. Johanna 7 June 1881. Signed by Sultan Abdallah, King of Johanna and G. Succi': ibid., 29.
14 'L'isola che più di tutte, e fino al primo viaggio, attire la mia attenzione, e per la ricchezza de'suoi prodotti e per la centrica posizione nell'Arcipelago, fu l'isola di Johanna, la quale oggidi acquisterebbe una vitale impotanza pel commercion, avendo potutto ottenere de quel Re una piena perpetua franchigia per tutti gli articoli d'importazione e d'esportazione': ibid., 4.
15 'Stanley reached London on the day that Succi, the wonder of the time, finished his great 40 days fast, and arrived in New York harbour on the day that Succi began his 45-day attempt in New York': *Pall Mall Gazette*, 26 Feb. 1890, 2; 'Empty Bread Baskets as a Steady Diet'; Succi, *Commercio in Africa*, 29; Eusebio Martínez de Velasco, 'El viajero italiano Juan Succi en el día vigesimoctavo de su ayuno', *La Ilustración Española y Americana*, 36 (1886), 182.
16 Giovanni Chiverny, *Del Signor Succi e del suo digiuno* (Milan: Tipografia di Giacomo Pirola, 1886).
17 'Il rappelle qu'Hippocrate a été le témoin d'un jeûne volontaire de quinze jours. De même, Galien. Il ne parle que pour mémoire du jeûne des saints et des saintes. En 1603, à Pise, l'Université a suivi le jeûne d'une jeune fille qui a été privée de nourriture pendant quinze mois. En 1618, un Provençal est resté huit mois sans manger. A Brunswick, une jeune fille a refusé toute boisson pendant six ans, etc., etc.': *L'impartial*, 12 Nov. 1886, 4.
18 Pietra Santa had close links with Ernest Monin, one of the promoters of Merlatti's fast in Paris in 1886. See for example Ernest Monin, *La crémation. Avec lettre-préface du docteur Prosper de Pietra Santa* (Paris: L'hygiène pratique, 1883).
19 *Pall Mall Gazette*, 26 Feb. 1890, 2 (my emphasis).
20 Chiverny, *Del Signor Succi e del suo digiuno*, 22.
21 Robida's satirical approach to Succi's travels in Africa appeared as 'Aventures du célèbre jeûneur Luigi Macaroni, par Georgina', *La Caricature*, 25 Dec. 1886, 421. The original French captions read:

> Transporté par le vapeur la Méduse, dans un des coins les plus reculés de l'Océan, Luigi Macaroni se laisse emporter au hasard des vents.
> 60e jour. Per Bacco ! Ze prouverai bien que le zeune et l'eau d'Hunyadi Janos sont les plus grandissimes conquêtes de la science !

180e jour. Par la Madone ! Z'aurais dû prendre un pardessus !
275e jour. Les Zéphirs ont transporté le signor Macaroni dans l'ile Thibia-grati-nophom habitée par la tribu des Bouff-lé-néh-soss-tomath, anthrophages de profession.
Les Bouff-lé-néh-soss-tomath sont des gourmets qui méprisent un si maigre régal. Macaroni en est profondement humilié.
Le roi des animaux lui-même passe dédaigneux. Macaroni est plus en plus humilié.
Tu ne veux pas manzer? eh bien! ze te manzerai!
365e jour. Retour à Paris. Un lion pommes purée ! boum voilà!'

22 'Questo digiuno, che fu compiuto al Royal Acquarium, Westminster, destò un interesse grandissimo negli scienziati e nel pubblico, in generale, e l'Acquarium era ogni giorno affollatissimo. Medici e alter convennero da ogni parte dello estate per esaminare e conversare con un uomo cosí straordinario, e a la stampa londinese e della provincial dedicò grandi colonne sull'andamento dell'esperienza': *Cenni Biografici. Giovanni Succi. Esploratore d'Africa. Già delegato della 'Società di Commercio coll'Africa' di Milano e membro della 'Società psicologica di Madrid'* (Rome: Tipografia Económica, 1897), 10.

23 Peter Payer, *Hungerkünstler in Wien. Eine verschwundene Attraktion* (Vienna: Verlag Sonderzahl, 2002).

24 'Para ayer sábado a las cinco de la tarde, estaba anunciado el ingreso en una caseta, construida en el pórtico del Teatro Principal, del individuo llamado Juan Succi, que dice permanecerá allí incomunicado y sin tomar ninguna clase de alimento durante 30 días. Con motivo de esta sensacional experiencia, *que se anuncia como científica*, un Periódico de ayer pidió al coronel Félix Díaz, Inspector General de Policía, *y con objeto de evitar la burla del público*, se sirva establecer un Servicio de policía, que garantice el exacto cumplimiento del compromiso contraído por el ayunador. Daremos cuenta de lo que suceda': *El Tiempo*, 30 Apr. 1905, 2.

25 There is evidence of some impostors using his name in Buenos Aires: '¿Qué pasará con Succi? ... A propósito de este ayunador, recordemos que las autoridades de Buenos Aires, allá por los años 1900 o 1901, tomaron y redujeron a prisión a un individuo del mismo nombre, que se hacía pasar por ayunador' (*La Patria de México*, 19 May 1905, 1). The *New York Times* also referred to a Succi committing fraud in Rio de Janeiro in 1899: 'Succi, the famous Italian faster, has been unmasked here. Dr Daniel Almeida has discovered that he used fibrous meat compressed into the smallest size. This, with a small quantity of mineral water, was enough to prevent starvation' ('Succi, the Faster, Exposed. Found That He Used Fibrous Meat and Mineral Water', *New York Times*, 10 Dec. 1899, 7).

26 'Le jeûneur Succi et sa femme sont en ce moment à Nancy, où ils se proposent de donner une série d'attractions': *La Press*, 27 Feb. 1900, 3.

27 For Merlatti's fast in London in 1885, see *Illustrated London News*, 91 (1887), 88; Nieto-Galan, 'Useful Charlatans'.

28 '[J]e vis Succi ingénieur, architecte, construisant des deniers de son jeûne une vaste maison, sorte de petit village, pour nourrir à des frais les affamés d'Italie': Ernest Monin and Philippe Maréchal, *Stefano Merlatti. Histoire d'un jeûne célèbre, précédée d'une étude anecdotique, physiologique et médicale sur le jeûne et les jeûneurs* (Paris: C. Marpon et E. Flammarion, 1887), 109–110.

29 The press reported the event in the following terms: 'Berlin has now started its fasting man, who is called Franceso Cetti. He has commenced a fast, which is to last 30 days. Many journalists, physicians and others attended the commencement of his undertaking. Cetti is a sickly-looking man of 27 ... During the fast he is to take nothing but *distilled water and Vichy water*. Two physicians will watch him day and night, and two professors have undertaken the scientific experiments and observations' (*Yorkshire Post and Leeds Intelligencer*, 14 Mar. 1887, 5), https://digitaltmuseum.no/011014280882/cetti-i-sin-ballon (accessed 10 Oct. 2018).
30 Wilfrid de Fonvielle, *Histoire de la navigation aérienne* (Paris: Hachette, 1911).
31 *Norsk Biografisk Leksikon*, https://nbl.snl.no/Francesco_Cetti (accessed 21 Jun. 2020).
32 Such as Curt Lehmann, Müller, Munck, Senator, and Zuntz: *Archiv für pathologische Anatomie und Physiologie und für klinische Medicine*, 13 (1893), 131, Supp., 1–228.
33 'He is 45 years of age, and served in the Franco-German War ... Jacques, who is about 5 ft. 5 in. in height, "plump", but not stout, has declared that last spring he fasted 28 days, his only sustenance being his elixir, prepared from herbs which may be gathered on Dartford Heath': *The People*, 07 Nov. 1886, front page.
34 Jacques's fast was later subjected to a medical experiment by Drs Paton and Stockman: Paton and Stockman, *Proceedings of the Royal Society of Edinburgh*, 16 (1888–1889), 121.
35 'Emigrants of Doubtful Utility: Dime Museum Cranks Whose Exhibitions Have No Value and Who Are Utterly Worthless as Citizens', *Chattanooga Daily Times*, 30 May 1892, 2.
36 *Norsk Biografisk Leksikon*, https://nbl.snl.no/Francesco_Cetti (accessed 21 Jun. 2020).
37 'Sacco Homan. The Fasting Man', *Register* (Adelaide), 19 Mar. 1910, 3; https://blogs.slv.vic.gov.au/such-was-life/the-marvellous-and-macabre-waxworks-part-one-the-more-murderers-the-more-it-thrives/ (accessed 2 Jun. 2018). See also https://oztypewriter.blogspot.com/2020/11/hunger-games-have-typewriter-will-fast.html (accessed 2 Jun. 2018).
38 'Fasting Man in Ipswich', *Framlingham Weekly News*, 24 Feb. 1912, 4.
39 Huber and Osterhammel, *Global Publics*.
40 Sebastian Conrad, *What Is Global History?* (Princeton, N.J.: Princeton University Press, 2016), 97.
41 Winchester, *Krakatoa*.
42 Conrad, *What Is Global History?*
43 Osterhammel and Camiller, *The Transformation of the World*.
44 Léonard Laborie, 'Global Commerce in Small Boxes: Parcel Post, 1878–1913', *Journal of Global History*, 10 (2015), 235–258; Aiqun Hu, 'The Global Insurance Movement since the 1880s', *Journal of Global History*, 5 (2010), 125–148.
45 'A global public is a very large group, dispersed transnationally and, mostly, transcontinentally. Its members are, as a general rule, unknown to each other, but share a common focus of attention': Huber and Osterhammel, *Global Publics*, 16–17.

46 On concepts such as synchronicity and micro-histories of the global, see Conrad, *What Is Global History?*

47 A letter from Tanner to Linda Hazzard on 23 February 1912: Linda B. Hazzard, *Scientific Fasting* (New York: Grant Publications, 1927), 16.

48 '[N]ous assistons depuis quelque temps à un défilé de docteurs excentriques bien fait pour réjouir … Nous avons d'abord le docteur Tanner, le maître des maîtres, qui a essayé de prouver ex professo que l'homme n'est pas né pour manger, et que les deux repas par jour sont un préjugé à extirper au plus vite': *La Caricature*, 29 Jan. 1881, 35. At the end of his life, Tanner pointed out that '[his] novel experience of fasting forty days created a profound sensation at the end of the telegraphic world' (Henry S. Tanner, *The Human Body: A Volume of Divine Revelations, Governed by Laws of God's Ordaining, Equally with Those Written on 'Tables of Stone' or in the Bible* (Long Beach, Calif., 1908), 7).

49 'Ahora se ha descubierto que el doctor Tanner tomaba por las noches, durante su célebre ayuno, extracto de carne Liebig': *El Loro*, 04 Sep. 1880; William H. Brock, *Justus von Liebig: The Chemical Gatekeeper* (Cambridge: Cambridge University Press, 2002).

50 'Il n'est pas de bonne philosophie de nier a priori, et de se refuser obstinément à étudier une question qui a passionné à tout l'Europe et même le monde entier': L'abbé Vallée, curé de Monts, 'Essai d'explication théorique et scientifique du jeûne extraordinaire des italiens Succi et Merlatti', *Annales de la Société d'agriculture, sciences, arts et belles lettres d'Indre-et-Loire*, 67 (1887), 29.

51 The *Argus* reported that, at the age of twelve, Merlatti had tried fasting. Before Paris, he had also fasted for eighteen days in Turin and later for thirty-six days in London: *Argus*, 11 Dec. 1886.

52 Mr Simon, Brussels, 1886, www.sideshowworld.com/13-TGOD/2014/Hunger/Artists.html (accessed 20 Dec. 2020). In this case, however, not much is known about Simon's biography. Several consistent patterns with other fasters such as Tanner, Succi, and Merlatti are evident: he was supervised by a medical board; he triggered a debate on the limits of human inanition; he was interested in financial gain; and he justified the oddity of his act with moral arguments.

53 *Toronto Daily Mail*, 9 Dec. 1886.

54 'He avers that after his last fast of twenty-eight days he was able to go to work four days after he concluded his fast': *The People*, 7 Nov. 1886.

55 *Suffolk and Essex Free Press*, 8 Dec. 1886, 3.

56 'Jeûneur, euse', *Grand dictionnaire universel du XIXe siècle* (Paris: Pierre Larrousse, 1866–1877), 17 (2), 1459.

57 *Cenni Biografici. Giovanni Succi*, 3.

58 'Considerable interest has been aroused in the scientific world in Rome by a gentleman named G. Succi, who professes to have discovered a liquor a small quantity of which enables a man to fast for 30 days, or even two months at a time': *Notre Dame Scholastic*, 20 (5) (1886), 79.

59 'A preuve que la faim est une sensation purement nerveuse, nous citerons, les fumeurs, qui mangent peu parce qu'ils se sont narcotisé l'estomac … On peut également apaiser la faim par l'opium (les mangeurs ou fumeurs d'opium ne font usage d'aucune nourriture, il est vrai qu'ils sont émaciés). C'est plutôt une prostration nerveuse. Or, en faisant boire à des chiens

soumis à une abstinence prolongée, de l'eau dans laquelle on mêle une certaine quantité de strychnine, on empêche, ou du moins on retarde cette prostration': Adolphe Burggraeve, *Longévité humaine, par la médecine dosimétrique ou la médecine dosimétrique à la portée de tout le monde, avec ses applications à nos races domestiques* (Paris: Librairie et gare de chemin de fer, 1887), 197.

60 'Miscellaneous News', *Poverty Bay Herald*, 20 Dec. 1888, 3. See also 'To Fast Fifty Days: Stefano Merlatti's Attempt in Paris and a Talk with Him', *Columbus Daily Enquirer*, 13 Nov. 1886.

61 *Victoria Daily Colonist*, 12 Nov. 1890.

62 George M. Gould and Walter L. Pyle, *Anomalies and Curiosities of Medicine* (Philadelphia: W. B. Saunders, 1897), 421; Payer, *Hungerkünstler in Wien*.

63 'En 1909, il y eut deux nouvelles expériences de jeûne faites dans un but scientifique: l'une en Angleterre, l'autre en Allemagne. M. Penny, médecin anglais, s'est amusé à jeûner trente jours durant. Ce n'était point pour gagner un pari, ou pour s'exhiber à, ses contemporains: il jeûna chez lui tout simplement, sans la moindre ostentation et pour voir de combien la privation d'aliments le ferait maigrir et diminuer de poids. Tout le long de l'expérience, M. Penny observa son pouls, son poids, sa respiration et examina son sang. Le jeûne fut absolu: le sujet ne prenant que de l'eau distillée. Il occupa son temps à lire, à converser et à faire de l'exercice. Il passait de douze à quatorze heures au lit. L'exercice consistait en promenades à pied et à bicyclette: la marche était en moyenne de cinq kilomètres et demi; la course à bicyclette de huit kilomètres et demi. M. Penny eut faim pendant les deux premiers jours seulement; après quoi cette sensation lui passa. Ce dont il s'est plaint le plus, c'est d'avoir froid, surtout aux pieds et aux mains. Le trentième jour, le jeûne prit fin: le sujet absorba une livre (livre anglaise de 453 grammes) de fruits, ce qui ne l'empêcha pas de perdre encore une livre de poids en dix-sept heures, en même temps qu'il se faisait une abondante expulsion d'urates. La perte de poids du début à la fin de l'expérience avait été de 13 k. 137 grammes, soit de 438 grammes par jour': Albert de Rochas, *La suspension de la vie* (Paris: Dorbon-ainé, 1913), 33–34.

64 Ibid., 35.

65 'Reuters Memorandum to Correspondents (1883)', Reuters Archive. I am indebted to Mr Rory Caruthers, at the Reuters Archive (London) for sending me this document.

66 Johannes Müller (1801–1858) was a German physiologist and the author of the widely read *Handbuch der Physiologie des Menschen für Vorlesungen*, 2 vols. (Coblenz: Verlag von J. Hölscher, 1837).

67 Nathan Zuntz (1847–1920) was a German physiologist interested in metabolism, respiration, nutrition, and aviation medicine. In collaboration with Curt Lehmann, he was involved in experiments with hunger artists such as Cetti and Breithaupt.

68 Brock, *Justus von Liebig*.

69 Robert G. Frank Jnr, 'American Physiologists in German Laboratories, 1865–1914', in Gerald Geison (ed.), *Physiology in the American Context, 1850–1940* (Bethesda, Md.: American Physiological Society, 1987), 11–46.

70 Gerald L. Geison, *Michael Foster and the Cambridge School of Physiology: The Scientific Enterprise in Late Victorian Society* (Princeton, N.J.: Princeton University Press, 1978).
71 Geison (ed.), *Physiology in the American Context.*
72 '[V]enu exprès pour rendre compte de son observation. Le docteur Thomas Linn a été des membres assistants au jeûne célèbre du docteur Tanner. Il a trouvé que Merlatti présentait ou 18e jour les mêmes phénomènes que ceux observés chez son confrère américain': Monin and Maréchal, *Stefano Merlatti*, 193–194.
73 '[D]octeur Chassaing, conseiller municipal de Paris; docteur Baraduc, de Châtel-Guyon; docteur Marcellin-Cazaux; docteur Cl. Clement; docteur R. De Lagenhaguen; docteur Camussi; Louis Figuier; docteur Wilt; docteur de Wecker, etc.': ibid., 225.
74 With his strong emphasis on experimentation and measurements, Moleschott had a profound influence on Italian physiology, but he was also interested in more theoretical and philosophical approaches.
75 Essays in honour of Luciani also received contributions from international luminaries in the discipline such as Charles Richet (Paris), M. Benedict (Vienna), J. N. Langley (Cambridge), and Paul Heger (Brussels).
76 Agustí Nieto-Galan, 'Hunger Artists and Experimental Physiology in the Late Nineteenth Century: Mr Giovanni Succi Meets Dr Luigi Luciani in Florence', *Social History of Medicine*, 28 (1) (2015), 82–107.
77 *British Medical Journal*, 21 Jun. 1890, 1444–1446.
78 During Succi's fast in Florence in 1888, Luciani had a huge network of collaborators who represented the Italian web of physiology of his time. Dr Virgilio Ducceschi was Fano's pupil at the physiology laboratory in Florence; Dr G. Buffalini was the director of the Gabinete di Materia Médica and Farmacologia; Dr Dario Baldi was at that time professor of pharmacology in Pisa, but collaborated with Luciani as a supervisor of the experiment on Succi in Florence; Dr E. Tanzi was the director of a psychiatric clinic, also in Florence; Dr Amedeo Herlitzka had followed the Moleschott–Mosso connection at the laboratory of physiology in Turin; Dr Luigi Tarulli was at the physiology laboratory of the University of Rome. All were involved in tests and measurements during Succi's fast. The whole team measured: urine analysis (Dr Baldi, Dr Roster, Dr Pons, Dr Pelizzari); soluble ferments of saliva, vomit, and urine (Dr Luciani); gaseous exchange through the lungs (Dr Luciani); anthropometrical examination (Dr Bianchi); weight and muscular strength (Dr Luciani); blood analysis (Dr Colzi); bacteriology (Dr Banti); the senses (Dr Silvestri and Dr Luciani); psychic examination (Dr Filippi); temperature and pulse (Mrs Burresi, Pieraccini, Favilli, and medical students): 'Succi, The Fasting Man', *The Lancet*, 17 Mar. 1888, 53.
79 Neswald, 'Strategies of International Community-Building'.
80 Ibid.
81 Ibid., 12.
82 Francis Gano Benedict, *A Study of Prolonged Fasting* (Washington, D.C.: Carnegie Institution of Washington, 1915).

83 London 1890: *British Medical Journal*; New York 1890: *New York Daily Tribune*; Naples 1892: Drs Ajello and Solaro, and Drs Dutto and Lo Monaco; Vienna 1896: Drs E. and O. Freund; Zurich 1896: Dr Daiber; Hamburg 1904: Dr Brugsch. See Benedict, *A Study of Prolonged Fasting*, 16–18.
84 Francis Gano Benedict, *The Influence of Inanition on Metabolism* (Washington, D.C.: Carnegie Institution of Washington, 1907), 7.
85 US Department of Agriculture, *Experiment Station Record*, 19 (1897–1898), 202; Josef Brozek, 'Six Recent Additions to the History of Physiology in the USSR', *Journal of the History of Biology*, 6 (2) (1973), 317–334.
86 Sergius Morgulis, *Fasting and Undernutrition: A Biological and Sociological Study of Inanition* (New York: Dutton and Company, 1923), 90, 153.
87 Benedict, *The Influence of Inanition on Metabolism*, 1.
88 'Les jeûneurs devant la science', *L'impartiel*, 12 Nov. 1886, 1; Luciani, *Fisiologia dei digiuno*, 24.
89 The report pointed out that Succi's case might be of scientific interest and could draw the attention of professional doctors if he could provide new data on the physiology of nutrition and fasting. 'Solo en este caso estaría justificado que profesores serios y que estimen algo su prestigio, acudiesen a perseguir durante treinta días el ayuno y los ejercicios del conocido italiano. De no ser así, más vale que la clase médica no intervenga en un acto que ha sido fiscalizado en otras muchas poblaciones y que en último extremo no representaría hoy de positivo más que aumentar su solemnidad en beneficio de los señores Ducazcal y Succi': Decio Carlán, 'El ayunador Succi. La Sección de Ciencias Naturales del Ateneo', *El Siglo Médico*, 9 Dec. 1888, 785.
90 Ricci was also the manager of *Il Corriere Spiritico*, the spiritist journal edited by Succi in Florence. Again stressing the commercial nature of the whole endeavour, Ricci talked about Succi in the following terms: 'Quando Succi digiuna, io mangio. Quando Succi mangia, digiuniamo tutti e due!' ('When Succi fasts, I eat. When Succi eats, we both fast!') (Sistema Museale Provincia di Ravena, www.sistemamusei.ra.it/main/index.php?id_pag=99&op=lrs&id_riv_articolo=514 (accessed 23 Jun. 2018)).
91 'Succi Wins His Race with Starvation', *New York Herald*, 21 Dec. 1890.
92 'George Francis Train Starving Himself to Death', *Dundee Courier*, 7 Jan. 1878, 4.
93 'Ups and Downs of Succi's Spell', *New York Herald*, 21 Dec. 1890. As reported in the press, Succi and Train met at the Dime Museum in Boston in 1890: 'Succi, the great Italian long distance faster, and George Francis Train were today the attraction of the Nickelodeon, a dime museum here, the one as a freak, the other as an accompanist. Arm in arm they trod hourly the stage, George Francis Train with the air of an entrepreneur, Succi with the dejected bearing of one of Stanley's rear guard. Their appearance was the signal for vociferous applause. Several thousand people paid to see the couple today' ('Succi and George Francis Train: Crowds Visit a Dime Museum to See the Couple', *Chicago Herald*, 21 Dec. 1890).
94 Giovanni Corvetto, 'I ricordi di un impresario teatrale. Succi e i suoi fenomenale digiuni', *La Stampa della Sera*, 29 Mar. 1932, 3.

2 Performances

1 Maud Ellmann, *The Hunger Artists: Starving, Writing and Imprisonment* (London: Virago Press, 1993), 17.
2 Thomas A. Markus, *Buildings and Power: Freedom and Control in the Origin of Modern Building Types* (London: Routledge, 1993).
3 Aileen Fyfe and Bernard Lightman (eds.), *Science in the Marketplace: Nineteenth-Century Sites and Experiences* (Chicago: University of Chicago Press, 2007).
4 Not much is known about hunger artists' performances in international exhibitions: Guido Abbatista (ed.), *Moving Bodies, Displaying Nations: National Cultures, Race and Gender in World Expositions Nineteenth to Twenty-First Century* (Trieste: Edizioni Universita di Trieste, 2014).
5 Barbara Gronau, 'Asceticism Poses a Threat: The Enactment of Voluntary Hunger', in Alice Lagaay and Michael Lorber (eds.), *Destruction in the Performative* (Amsterdam: Rodopi, 2012), 99–109.
6 Kafka, *A Hunger Artist*, 58–59. I discuss this in Agustí Nieto-Galan, 'Useful Charlatans: The Fasting Contest of Giovanni Succi and Stefano Merlatti in Paris, 1886', *Science in Context*, 33 (4) (2020), 405–422.
7 Ernest Monin and Philippe Maréchal, *Stefano Merlatti. Histoire d'un jeûne célèbre, précédée d'une étude anecdotique, physiologique et médicale sur le jeûne et les jeûneurs* (Paris: C. Marpon et E. Flammarion, 1887), 154.
8 'Aperitivo Xerez Quina Ruiz, Consommé, Poisou Frit [*sic*], Bœuf Braisé, Poulet Roti, Dessert: Café. Champagne Moët Chandon, Whit [*sic*] Seal, Cognac B. Vert, Benedictino, Cigarros Buen Tono. Aguas de mesa: Cruz Roja Tehuacán. Entrada: domingos, 50 centavos': *El Imparcial*, 19 May 1905.
9 *Illustrated Police News*, 26 Apr. 1890 and 3 May 1890.
10 David Livingstone, *Putting Science in Its Place: Geographies of Scientific Knowledge* (Chicago: University of Chicago Press, 2003).
11 Agustí Nieto-Galan, *Science in the Public Sphere: A History of Lay Knowledge and Expertise* (London: Routledge, 2016), ch. 3, 'Spectacular Science'.
12 *Graphic*, 20 Sep. 1886.
13 'Succi vint ensuite à Paris. Quand il eut, à grand peine, après un mois de démarches, réussi à constituer un comité, son imprésario, le chevalier Lamperti, l'exhiba d'abord dans un appartement de la rue Le Peletier, avec un tourniquet. L'entrée coûtait 2 francs la semaine et 1 franc le dimanche; malgré la modicité des prix, il n'y eut presque pas de visiteurs. Le malheureux passa alors à l'état d'annexé dans des établissements comme l'Olympia, l'Eden-théâtre et les Montagnes russes, mais il n'eut pas plus de succès. C'est dans un de ces établissements que je l'ai vu et, comme j'étais à peu près seul avec lui, je pus causer assez longtemps. Il me parut très versé dans les sciences psychiques et d'un esprit parfaitement équilibré': Albert de Rochas, *La suspension de la vie* (Paris: Dorbon-ainé, 1913), 32. For rollercoasters, see Jaume Sastre-Juan and Jaume Valentines, 'Technological Fun: The Politics and Geographies of Amusement Parks', in Oliver Hochadel and Agustí Nieto-Galan (eds.), *Barcelona: An Urban History of Science and Modernity (1888–1929)* (London: Routledge, 2016), 92–112.

14 'Les médécines l'ont autorisé à faire une longue promenade à pied. Il ira du Grand Hotel ou Bois de Boulogne et reviendra. Une voiture suivra le groupe au pas, en cas d'accident': Monin and Maréchal, *Stefano Merlatti*, 190.

15 'Stefano Merlatti continue ses séances de peinture et du dessin. Il réçoit ses visiteurs de midi à six heures du soir. A six heures et demie, il sort faire sa promenade et rentre généralement de huit à neuf heures, toujours accompagné, toujours surveillé, toujours épié': ibid., 192.

16 Agustí Nieto-Galan, 'Hunger Artists and Experimental Physiology in the Late Nineteenth Century: Mr Giovanni Succi Meets Dr Luigi Luciani in Florence', *Social History of Medicine*, 28 (1) (2015), 82–107.

17 Cinematografo della Borsa, Bologna, 30 Sep. 1907: 'per pubblicizzare il locale ... l'invito del celebre digiunatore Succi. Fece construire nella salla d'aspetto una gabia di vetro porta su una pedana di legno di un metro d'altezza, e dentro vi collocò il Succi, che digiunò per un mese intero, alla presanza degli spettatori increduli': Ennio Ferretti, 'Giovanni Succi, il digiunatore di Cesenatico che ispirò Kafka', *SeiTorri*, 27 Feb. 2019, www.seitorri.it/giovanni-succi-il-digiunatore-di-cesenatico-che-ispiro-kafka/ (accessed 10 May 2020).

18 'Succi dans une boîte: un lit ... un thermomètre, un pair de ciseaux, un rasoir, un tire-bouchon, un morceau de savon, deux crayons, un carnet, un paquet de bougies, 7 bouteilles d'eau de Vichy, et un petit flacon de "ce fameux élixirs qu'il a inventé et que ne fit pas fortune", un bon roman français: Pot-Bouille de Zola, 50 centimes de franc l'entrée, gymnastique, l'escrime, équitation': Giorgio del Rio, 'Lettre de l'Italie', *Le Diplomate*, 18 Jul. 1897, 7.

19 Javier de Benavent, *Ayuno Succi* (Barcelona: Tipo-Litografía de Busquets y Vidal, 1890), 14.

20 Dr Avelino Martín described him as: 'Joven aún, de fino porte y agradables maneras se presenta, no sólo como un curioso fenómeno científico digno de serios estudios, sino que también por la singularidad de hecho que lleva a cabo se hace interesante aún a los más apáticos caracteres': Avelino Martín, *El ayuno de Succi. Contribución al estudio de la inanición* (Barcelona: Tipografía de la Casa Provincial de Caridad, 1889), 4.

21 Since the 20 per cent of the revenue from the poor reached 700 pesetas, we can conclude that the approximate figure of official visitors was 7,000: 'Succi en el Ateneo', *La Vanguardia*, 17 Nov. 1888, 1.

22 'Globo Cautivo, – Exposición Universal. – Ascensiones diarias, si el tiempo lo permite desde las 6 1/2 de la mañana á las 12, y de las 2 de la tarde al anochecer. Ascensiones nocturnas a precios convencionales. – Entrada al recinto del Globo, 0'50 pesetas. Billete de ascensión, 5 pesetas. NOTA: Recomienda la empresa al público aproveche de las 6 1/2 a las 12 de la mañana como las mejores horas para las ascensiones': *La Vanguardia*, 25 Jun. 1888, 2.

23 *La Correspondencia de España*, 14 Dec. 1888, front page. Succi later complained about the conditions at the Teatro Felipe, and wanted to move the fast to the Teatro Apollo.

24 *El Diario del hogar*, 8 Mar. 1905. I am indebted to Patricia Guerrero, who provided me all the details of Succi's stay in Mexico and references to Cuba and Argentina in the Mexican press.

25 *El Mundo*, 25 Apr. 1905.
26 *El Imparcial*, 28 Apr. 1905; *El Correo Español*, 1 May 1905.
27 *El Tiempo*, 13 May 1905.
28 'Aujourd'hui samedi, à dix heures du matin, le jeûneur Succi depuis quinze jours aux Montagnes russes, fera l'ascension de la tour Eiffel, accompagné d'une commission médicale': *Le petit journal*, 16 Jun. 1889.
29 'Ups and Downs of Succi's Spell', *New York Herald*, 21 Dec. 1890.
30 *Pall Mall Gazette*, 26 Feb. 1890, p. 2.
31 'Al termine di quest'esperimento a New York [1890], il Succi dedicò un gran period di tempo a viaggiare per l'America e il Canadà, e non ritornò in Europa fino alla Primavera': *Cenni Biografici. Giovanni Succi. Esploratore d'Africa. Già delegato della 'Società di Commercio coll'Africa' di Milano e membro della 'Società psicologica di Madrid'* (Rome: Tipografia Económica, 1897), 12.
32 *Le Rappel*, 19 Apr. 1891. George Bunnell, P. T. Barnum's protégé, played a key role in the development of the American dime museum to increase the number of visitors and allow the lower classes to access the oddities: https://travsd.wordpress.com/2013/04/06/george-b-bunnell/ (accessed 22 Jun. 2020).
33 'Clinton, Mo., Jan. 11. – Dr. Tanner, the man who was famous so long for having fasted forty days, now lives on a farm eighteen miles southwest of this city. He now challenges Signor Succi to sit down with him in Chicago during the World's Fair to test the matter in a ninety days' fast on water only, or if Succi prefers, to let the fast continue from day to day till one or the other yields the contest': *New York Times*, 12 Jan. 1891.
34 *Cenni Biografici. Giovanni Succi.*
35 Peter Payer, *Hungerkünstler in Wien. Eine verschwundene Attraktion* (Vienna: Verlag Sonderzahl, 2002).
36 Peter Payer, 'Hungerkünstler. Anthropologisches Experiment und modische Sensation', in Brigitte Felderer and Ernst Strouhal (eds.), *Rare Künste. Zur Kultur- und Mediengeschichte der Zauberkunst* (Vienna: Springer, 2006), 254–268.
37 *San Francisco Chronicle*, 25 May 1905, 13.
38 Ibid.
39 See www.spiegel.de/fotostrecke/hungern-als-show-fotostrecke-107202.html (accessed 15 Jun. 2021).
40 'The Fasting Man', *Illustrated London News*, 26 Apr. 1890, 516.
41 Ibid.
42 Ibid.
43 Ibid.
44 For example, during Merlatti's fasts in Paris in 1886.
45 In 1900 in Madrid, the local press described Pappus's performance in the following terms: 'El distinguido y soñador artista que intentó dormirse durante varios días en el Teatro Romea y que desistió de sus propósitos de descanso a consecuencia de las informalidades de la empresa que le contrató, debutará en breve en Colón. M. Pappus es un hombre práctico; ha resuelto un difícil problema, vivir sin comer, y en esta gran habilidad suya, vive. Pappus es hombre de gran estatura, delgado (y es natural dado su género de alimentación), la fisionomía inteligente y ojos vivos. En sus viajes por toda Europa

ha suscitado gran curiosidad; no se ve todos los días a un hombre dentro de una urna como cualquier jarrón de flores artificiales o como Niño Jesús con peluca rubia': *El País*, 26 Aug. 1900, 2.

46 'Man in a Bottle: South American Lives for Eight Days Sealed Up in One', *Washington Post*, 15 Jan. 1905, 7.
47 *Pall Mall Gazette*, 26 Feb. 1890, 2.
48 Livingstone, *Putting Science in Its Place*, 1.
49 *Decatur Herald*, 5 Mar. 1905, 6.
50 Robert Bogdan, *Freak Shows: Presenting Human Oddities for Amusement and Profit* (Chicago: University of Chicago Press, 1988), 8–10.
51 Ibid., 37.
52 Ibid., 7.
53 John M. Munro, *The Royal Aquarium: Failure of a Victorian Compromise* (Beirut: American University of Beirut, 1971). See also Richard Altick, *The Shows of London* (Cambridge, Mass.: Belknap Press, 1978), 506; Bodleian Libraries, University of Oxford, John Johnson Collection.
54 Quoted in John Sands, 'Sullivan and the Royal Aquarium', Gilbert and Sullivan Archive. Arthur Sullivan was directly involved in the early design of the Royal Aquarium. See https://gsarchive.net/ (accessed 2 Jun. 2016).
55 Such as Ella Zuila and Lulu, the heroine of the lofty wire, the beautiful Madame Alphonsine, The Two Macs, Fish and Ralston, Lily Landon, The Tissots, Crossley and Elder, and the Wonderful Pole Leapers, among many others.
56 Erroll Sherson, *London's Lost Theatres of the Nineteenth Century* (London: John Lane, 1925), 297 (my emphasis); quoted in John Sands, 'Sullivan and the Royal Aquarium', https://gsarchive.net/ (accessed 2 Jun. 2016).
57 As described in the *New York Herald* by its London correspondent: 'At the end of ten days Succi was feeling better apparently than some of his newspaper critics … While many of the minor papers and some of the strong ones were deploring the exhibition and saying that it had no practical value, the medical journals appeared to treat it both as a serious and important matter': 'Empty Bread Baskets as a Steady Diet', *New York Herald*, 9 Nov. 1890.
58 *Globe*, 17 Dec. 1891, 2.
59 Bodleian Libraries, University of Oxford, John Johnson Collection: Entertainments Folder 13 (17).
60 'Physicians, Nobility, and the Press only – total about 1000 persons. Mr Cetti is the same faster who made this experiment before in the Berlin Medical Congress, under the superintendence of Gehaimrath, Professor Dr Virchow; and also here in London Mr Cetti has secured the leading physicians for his experiment. Mr Cetti will take his last dinner 12 o'clock midday. Beginning of the 30 days fasting, for which occasion the first physician will say a few words of the benefit done through this experience to medical science': *Morning Post*, 11 Jun. 1887, 1.
61 Pascal Blanchard, Gilles Boetsch, and Nanette Jacomijn Snoep, *Human Zoos: The Invention of the Savage* (Arles: Actes Sud, 2011).
62 Bogdan, *Freak Show*, 56.

63 In 1860, Barnum's American Museum on Broadway, New York, had exhibited thirteen 'human curiosities': an albino family, living Aztecs, dwarfs, a black mother with albino children, a bearded lady, fat boys, and a black man with intellectual disabilities. For more details on Barnum's freak culture, see Bogdan, *Freak Show*, 32–35.
64 *Le Rappel*, 19 Apr. 1891.
65 Jan Bondeson, *The Lion Boy and Other Medical Curiosities* (Stroud: Amberley Publishing, 2018), ch. 7, 'Fasting Artists', 110–121.
66 See National Fairground and Circus Archive, www.sheffield.ac.uk/nfca/projects/sheffieldjungle (accessed 15 Nov. 2020).
67 *Stonehaven Journal*, 8 Nov. 1906, 3.
68 *World's Fair Newspaper*, 6 May 1911, National Fairground and Circus Archive, Sheffield, www.sheffield.ac.uk/nfca/researchandarticles (my emphasis) (accessed 13 Sep. 2019). The daily press provided further details of Beauté's performance: 'The chief attractions this week at the Jungle are M. Victor Beauté, the Swiss fasting man, who commenced a month's abstention from food last Saturday, and Osco and the snakes. Mr Beauté accomplished his fast some eight years ago, when he went without food for 13 days. His idea is to benefit his health – he suffered from asthma – and after repeating the experience a time or two, he became completely cured. Mr Beauté says he has never felt any ill effects from his periods of fasting. During his Sheffield engagement his diet will consist of mineral waters (about two pints a day) and a dozen or so cigarettes a day': *Sheffield Daily Telegraph*, 4 May 1911, 4.
69 Bogdan, *Freak Show*, 1–21.
70 See https://travsd.wordpress.com/2011/05/21/variety-arts-4-dime-museums-and-side-shows/ (accessed 2 Jun. 2018).
71 Gronau, 'Asceticism Poses a Threat'.
72 Robert A. Gunn, *40 Days without Food. A Biography of Henry S. Tanner MD, including a Complete and Accurate History of His Wonderful Fast, 42 Days in Minneapolis, and 40 Days in New York City, with Valuable Deductions* (New York: Albert Metz, 1880).
73 Ibid., 16.
74 DAR, *Da Mosè a Succi. Storia e fisiologia del digiuno* (Florence: Fieramosca, 1886), 39.
75 Such as the dynamometer (strength), the esthesiometer (sensitivity of the nerves), and the sphygmograph (pulse).
76 Variables on the fortieth day of the fast were: weight loss (36 pounds), total amount of water drunk (681 ounces), total amount of urine passed (738 ounces); highest pulse (116), lowest pulse (66); highest temperature (100 °F), lowest temperature (97 °F); greatest strength (196 pounds), lowest strength (156 pounds). Dr Philip van der Weyde, professor of chemistry at the US Medical College, oversaw the quantitative measures of Tanner's metabolism: Gunn, *40 Days without Food*, 78–79, 83.
77 Ibid., 62.
78 *The Human Wonder of the World. The Celebrated Dr Tanner. A Full Account of His Forty Days Fast* (Philadelphia: Barclay and Co., 1880), 24.

79 Ibid.
80 Ibid.
81 'Like the Russian Nihilists, Dr Tanner repudiates the present science, the present religion, and the present social order. A regular physician, he repudiates the teachings of his own school; and in order to prove the correctness of his revolutionary medical opinions he has undertaken this heroic experiment ... As to religion and social order ... what he wants is the restoration of the "original primitive Christianity" ... in order to bring about a new social order, when there shall be no need of any prison, asylums, or poorhouses, since there will be a universal Christian brotherhood': ibid., 51–52.
82 Ibid., 29.
83 Ibid., 54.
84 Alexander Hermann (1844–1896), known as 'Hermann de Great', was an itinerant magician. Harry Houdini (1874–1926) – Erik Weisz – was a very popular magician, with an interest in spiritism and aviation.
85 *Tribune*, 10 Jul. 1880; quoted in Gunn, *40 Days without Food*, 98–100 (my emphasis). Just a year later, in 1881, in Chicago, John Griscom emulated Tanner and fasted for more than forty days, being supervised by Dr Leiter Curty through blood analysis. Although Griscom's achievement appeared in all the American press, some years later, in 1886, critics such as French populariser Wilfrid de Fonvielle denounced the inconsistency of red cell figures in the analysis, which could only be compatible with fraud: Wilfrid de Fonvielle, *Mort de faim, étude sur les nouveaux jeûneurs* (Paris: Librairie illustrée, 1886), ch. 6, 'Le contrôle du jeune'.
86 Monin and Maréchal, *Stefano Merlatti*, 103–145.
87 *British Medical Journal*, 21 Jun. 1890, 1444. 'Tanner's popularity increased. Each day he received an enormous amount of mail, women serenaded him at the piano and newspapers reported daily on his condition. He even received an offer from a Museum to stuff him if he should die': Walter Vandereycken and Ron van Deth, *From Fasting Saints to Anorexic Girls: The History of Self-Starvation* (London: Athlone Press, 1994), 84.
88 'Ups and Downs of Succi's Spell', *New York Herald*, 21 Dec. 1890.
89 Ibid.
90 '[Merlatti] seems reluctant to brave another 50 days of abstinence, and considers, shall we say, that the agent and the glory he has gained from it are not worth the trouble he has gone to': *Revue britannique*, 63 (1887), 266–267.
91 'Pourquoi prêter l'appui de noms respectés à une épreuve qui ressemble tant à une vulgaire mystification?': Paul Loye, 'La jeûne de Succi', *Bulletin du progrès médical*, 4 (1886), 862.
92 Paul Loye, 'Death by Decapitation', *American Journal of the Medical Sciences*, 97 (4) (1889), 387.
93 'Succi's Disgusting Exhibition [New York Telegram]', *New York Herald*, 20 Dec. 1890.
94 Luigi Luciani, *Fisiologia del digiuno: studi sull'uomo* (Florence: Tipografia dei Successori Le Monnier, 1889), 3–5.
95 'Succi: The Fasting Man', *The Lancet*, 10 Mar. 1888, 487.

96 Luigi Luciani, *Human Physiology*, 4 vols. (London: Macmillan and Co. Ltd, 1911), 421 (my emphasis).
97 Robert Fox, *The Savant and the State: Science and Cultural Politics in Nineteenth-Century France* (Baltimore: Johns Hopkins University Press, 2012).
98 Louis Figuier, 'Les nouveaux jeûneurs. Succi et Merlatti', *L'Année scientifique et industrielle*, 30 (1886), 402.
99 'On ne saurait prétendre, en résumé, que Succi et Merlatti soient des fous: ce sont toutefois des excentriques, des nerveux, des cérébraux, qui, soutenus par une foi exagérée dans leurs forces, suppriment chez eux la sensation de faim et résistent à l'inanition': ibid., 403.
100 Louis Figuier, *Le lendemain de la mort ou la vie future selon la science*, 11th ed. (Paris: Hachette, 1904; 1st ed. 1871).
101 Fonvielle published regularly in *Revue scientifique*, *La Nature*, and *Science illustrée*, and was the editor-in-chief of *L'Aérophile*. Fonvielle also published *Les saltimbanques de la science. Comment ils font des miracles* (Paris: M. Dreyfous, 1884).
102 '[N]ous ne croirons jamais à ce qui nous parait absurde à de qui révolte notre bon nom, et toujours nous insurgerons contre les apothicaires qui viendraient nous faire flairer l'haleine de Succi ou le selles de Merlatti, pour nous prouver qu'on peut jeûner 50 jours, sans d'autres réconfortants que l'autosuggestion et l'eau claire': Wilfrid de Fonvielle, *Les endormeurs. La verité sur les hypnotisants, les suggestionistes, les magnetiseurs, les donatistes, les braïdistes* (Paris: Librairie illustrée, 1887),14–15.
103 Fonvielle, *Mort de faim*, 297.
104 'On se dit qu'il faudrait que les gardes fussent bien plus que des médecines de plus haut mérite, ou des publicistes d'une réputation hors ligne pour avoir l'autorité morale nécessaire et contrecarrer les jugements de l'expérience constante de la raison. Ce seraient des anges du Seigneur descendues sur la terre, et nous des hommes exposés à toutes les erreurs humaines, que leur conviction ne s'imposera pas': Fonvielle, *Mort de faim*, 284.
105 'M. de Parville s'occupe, dans le Journal des Débats, de Succi et de Merlatti et fait à propos de leur jeûne prolongé, dont il ne conteste pas la réalité, des observations intéressantes, que nous reproduisons en partie. L'éminent vulgarisateur distingue tout d'abord entre l'inanition et la faim … Quoi qu'il en soit, conclut M. de Parville, et en résumé, il n'est pas douteux que, en dehors de tout état maladif, certaines personnes, soit par un simple effort de volonté, soit par éducation, puissent réellement résister à la faim et vivre encore assez longtemps aux dépens de leur propre substance. Il n'y a rien là qui soit extraordinaire non que renverse les lois de la physiologie': 'Le jeûne prolongé', *La revue des journaux et des livres*, 14 Nov. 1886, 57–58.
106 Aimée Boutin, 'Rethinking the Flâneur: Flânerie and the Senses', *Dix-Neuf*, 16 (2) (2017), 124–132; Gregory Shaya, 'The Flâneur, the Baduad, and the Making of a Mass Public in France, circa 1860–1910', *American Historical Review*, 109 (1) (2004), 41–77.
107 'Don Pedro se reunió con su mujer y sus hijos, que le esperaban, y con ellos fue a visitar al ayunador Succi, que ocupaba una habitación en el

propio Palacio de Ciencias. Allí estaba el émulo del Doctor norteamericano Tanner, custodiado y vigilado por una comisión facultativa. Oye tu, [Pedro], preguntó Doña Manolita a su marido: – ¿quién es ese sujeto?, – Es Succi, – ¿Chuchi? ... ¿Y qué hace?, – No come, – ¿Está malo?, – No, ya le ves, que lleva diez días sin comer, y anda, habla, brinca, maneja el florete, se ríe y no tiene novedad, – ¡Ave María! ¡qué guasonazo!, – Solo toma alguna purga que otra, y todos los días le lavan el estómago ..., – Pero puede que coma alguna cosita cuando nadie le vea, – No mujer, no come ¿No ves que ni en un momento le pierde de vista el consejo de vigilancia? ... – ¿Qué es lo que se propone este hombre sin comer?, – Pues, hija, dar de comer a su familia': Carlos Frontaura, *Barcelona en 1888 y París en 1889. Narraciones humorísticas* (Valencia: Pascual Aguilar, 1888), 65–66.

108 Gilbert Andrieu, *La gymnastique au XIXe siècle* (Joinville-le-Pont: Actio, 1999).

109 'Le jeûneur Succi', *Le petit parisienne*, 29 Nov. 1886, 2.

110 'Les jeûneurs devant la science', *L'impartiel*, 12 Nov. 1886, n. 1811, 1; Luciani, *Fisiologia dei digiuno*, 24.

111 'Le jeûneur Succi n'a vraiment pas de chance. Après s'être privé de nourriture pendant les trente jours convenus, avoir mis son estomac à la plus périlleuse des épreuves ... il espérait et avait le droit d'espérer recevoir le prix de son abstinence volontaire, c'est-à-dire quinze beaux mille francs que s'était engagé à lui payer, moyennant certaines conditions, un impresario, M. Lamperti': *L'Universe Illustré*, 15 Jan. 1887, 42.

112 Monin and Maréchal, *Stefano Merlatti*, 152–154.

113 See, for example, 'Après cinquante jours. La fin du jeûne de Merlatti', *Courrier des États-Unis*, 308 (1886), 1–3.

114 Joel Piqué, 'Procesos de construcción social y científica de la homeopatía en Catalunya (1890–1924)', PhD Thesis, Universitat Autònoma de Barcelona, Bellaterra, 2018, 42.

115 The composition of Succi's supervisory board was heterogeneous: Drs Salvador Badía, Javier de Benavent, Avelino Martín, Ramon Roig, Joan Pijoan, Joan Bassols, Julian Guerrero; ten journalists (A. Chiloni, L. Vera, C. Litrán, J. Critiés, among others); a medical surgeon, Dr Manuel Figuerola; nine medical students; two fencing teachers (S. Pardini, V. Scandolara); a photographer (Ginés Bonmatí); a translator (Luis Berjau); and three overseers. See Benavent, *Ayuno Succi*, 5–6. On 22 September 1888, Succi was ready to begin his public fast in a room in the Palacio de Ciencias (Science Pavilion) of the International Exhibition, in the Ciutadella Park. His health and behaviour were checked daily by three local boards of medical doctors, journalists and medical students, to make sure he was not ingesting any kind of nourishment. See *La Vanguardia,* 16 Sep. 1888, 2.

116 Gunn, *40 Days without Food.*

117 Monin and Maréchal, *Stefano Merlatti.*

118 *Le Rappel* covered Merlatti's fast daily.

119 *Manchester Courier*, 27 Dec. 1890.

120 Doctors Ingram, Mason, Wildman, Hagan, Bettini di Moise, Guiteras, Tyler, and Bauer constituted the medical board: *New York Herald*, 7 Nov. 1890.

121 Ibid.

122 'Is Succi's Secret Will Power or Paranoia?', *New York Herald*, 7 Nov. 1890.

123 'Is Succi's Secret Will Power or Paranoia?', 7 Nov. 1890; 'Empty Bread Baskets as a Steady Diet'; 'Succi Wins His Race with Starvation', 21 Dec. 1890; 'Doctors Wonder at the Succi Mystery', 8 Nov. 1890; 'Ups and Downs of Succi's Spell', 21 Dec. 1890: all in *New York Herald*.

124 'The Famous Faster: An Interview with Signor Succi', *Pall Mall Gazette*, 26 Feb. 1890, 2. In 1897, an anonymous biography of Succi describes his adventures in the following terms: 'Who is he and what is he? He is a marvel; the wonder of the nineteenth century. This is a bold phrase, but it comes from the *New York Herald*, and was said about Mr Giovanni Succi, African explorer and scientific faster, at the end of his last fast in New York. The *Herald* made this definition of Mr Succi as well as reporting on the conditions of his fast, reported with sworn testimony of his most trustworthy reporters … His conditions were minutely noted by the gentlemen at the *Herald*, and their correspondences occupied 143 columns of that newspaper … A short report of the world traveller, scientist and faster, on which the eyes of the civilised world are turned, forms an interesting story' ('Chi e che cosa è. Egli è una meraviglia; la mereviglia del Secolo decimonono. És questa una frase ardita, ma viene dal *New York Herald*, e fu detta riguardo al signore Giovanni Succi, esploratore africano e digiunatore scientifico, alla chiusa del suo ultimo digiuno a New York. L'*Herald* fece questa definizione del signor Succi come pure delle condizioni del suo digiuno, riportato con testimonianze giurate dei suoi *reporters* piu degni di fede … Le sue condizioni furono minutamente notate dai signori dell'*Herald*, e loro corrispondenze occuparono 143 colonne di quel giornale … Una breve rivista del mondiale viaggiatore, scienziato e digiunatore, su cui son rivoltu gli occhi del mondo civile, forma una storia interessante'): *Cenni Biografici. Giovanni Succi*, 3. The biography also cited its original sources: *New York Herald*, *Società Geografica Italiana*, *The Times*, *Daily Chronicle*, *Standard*, and other Italian newspapers.

125 *Cenni Biografici. Giovanni Succi*, 18.

126 John Rees, *Textbook of Medical Jurisprudence and Toxicology*, 4th ed. (Philadelphia: Blakiston's Son and Co., 1895), 188–191 (my emphasis).

127 John Pickstone, *Ways of Knowing: A New History of Science, Technology and Medicine* (Chicago: University of Chicago Press, 2000).

128 'As a professional faster, selling (via his impresario) the spectacle of his superhuman capacity to survive without food, he is a consummate figure of the changing cultural marketplace of the turn of the twentieth century': Alys Moody, *The Art of Hunger: Aesthetic Autonomy and the Afterlives of Modernism* (Oxford: Oxford University Press, 2018), 39.

129 Ibid., 40.

130 Virginia Nicholson, *Among the Bohemians: Experiments in Living 1900–1939* (London: Penguin, 2002); Nina Diezemann, *Die Kunst des Hungerns. Essstörungen in Literatur und Medizin um 1900* (Berlin: Kadmos, 2006); Moody, *The Art of Hunger*.

131 Francis G. Benedict, 'An Experiment on a Fasting Man', *Science*, 31 May 1912, 865.

132 'When a fast is made by a so-called "professional faster" for pecuniary gain, the subject is exhibited to the public as an attraction to the lovers of sensational

amusements. Three decades ago such exhibitions were not uncommon and in many instances the subjects consented (possibly in the hope of increasing the interest in their performance) to more or less strictly controlled observations of their fasts': Francis Gano Benedict, *A Study of Prolonged Fasting* (Washington, D.C.: Carnegie Institution of Washington, 1915), 11.

3 Experiments

1 Luigi Luciani, *Il cervelletto. Nuovi studi di fisiologia normale e patologica* (Florence: Coi tipi dei successori Le Monnier, 1891); Luigi Luciani, *Das Kleinhirn. Neue Studien zur normalen und pathologischen Physiologie* (Leipzig: Eduard Bosold, 1893). Later, Luciani developed his research with animals at the Laboratorio di Fisiologia del Istituto di Studi Superiori in Florence. See Agustí Nieto-Galan, 'Hunger Artists and Experimental Physiology in the Late Nineteenth Century: Mr Giovanni Succi Meets Dr Luigi Luciani in Florence', *Social History of Medicine*, 28 (1) (2014), 82–107; Domenico Priori, 'Il digiunanote e lo scienziato', *Rendiconti Accademia Nazionale delle Scienze detta dei XL. Memorie di Scienze Fisiche e Naturali*, 135 (2017), 1–9.

2 Agustí Nieto-Galan, 'Useful Charlatans: The Fasting Contest of Giovanni Succi and Stefano Merlatti in Paris, 1886', *Science in Context*, 33 (4) (2020), 405–422.

3 Ernest Monin and Philippe Maréchal, *Stefano Merlatti. Histoire d'un jeûne célèbre, précédée d'une étude anecdotique, physiologique et médicale sur le jeûne et les jeûneurs* (Paris: C. Marpon et E. Flammarion, 1887), 86; Jean-Vincent Laborde, *Biographie psychologique. Le cerveau, la parole, la fonction et l'organe, histoire authentique de la maladie et de la mort* (Paris: Schleicher Frères, 1898).

4 For animal experimentation and vivisection debates, see Nicolaas A. Rupke (ed.), *Vivisection in Historical Perspective* (London: Croom Helm, 1987); Anita Guerrini, *Experimenting with Humans and Animals: From Galen to Animal Rights* (Baltimore: Johns Hopkins University Press, 2003).

5 Charles Richet, 'Physiologie. Course de physiologie de la Faculté de Médecine de Paris. L'inanition chez les animaux', *Revue scientifique*, 21 (1889), 641–647; 23 (1889), 711–715; Charles Richet, 'Physiologie. Course de physiologie de la Faculté de Médecine de Paris. L'inanition chez l'homme', *Revue scientifique*, 4 (1889), 106–112; 26 (1889), 801–805.

6 Susan Lederer, *Subjected to Science: Human Experimentation in America before the Second World War* (Baltimore: Johns Hopkins University Press, 1995).

7 Charles Chossat, *Recherches expérimentales sur l'inanition. Mémoire auquel l'Académie des Science a décerné en 1841 le prix de physiologie expérimental* (Paris: Imprimerie royale, 1843).

8 'C'est precisement parce que l'expérience semblait bizarre et paradoxale qu'elle exigeat une surveillance plus attentive et plus sevère': Paul Loye, 'Le jeûne Succi', *Le progrès médical*, 4 (1886), 861.

9 George Hoggan, *Société française contre la vivisection fondée à Paris le 8 mai 1882. Extrait d'une lettre adressée au directeur du journal le 'Morning post', par le Dr George Hoggan, de Londres* (Paris: Imprimerie de P. Schmidt, 1883);

Daniel Metzer, *La vivisection. Ses dangers et ses crimes*, 2nd ed., *Notes et commentaires de Philippe Maréchal* (Paris: Au Siège sociale de la Société française contre la vivisection, 1906).

10 Pierre Gallois, 'Philippe Maréchal', *La presse médicale*, 10 Mar. 1926, 315–316.

11 'Les phases physiologiques et les péripéties du jeûne humain sont moins connues que dans les animaux ... Heureusement, *des jeûnes professionnels*, par appât de gain et de publicité, ont réussi à persister bien plus longtemps dans l'abstinence. Ils ont consenti à se prêter en même temps aux déterminations scientifiques nécessaires sur eux et leurs éliminations': Henri Labbé, 'L'inanition. Ses aspects physiologiques et sociaux', *Revue scientifique*, 18 (1908), 549 (from a lecture by Charles Richet at La Sorbonne; my emphasis); Henri Labbé, *Principes de la diéthethique moderne* (Paris: Baillière et Fils, 1904).

12 Lederer, *Subjected to Science*.

13 Elizabeth Neswald, 'Food Fights: Human Experiments in Late Nineteenth-Century Nutrition Physiology', *Clio Medica*, 95 (2016), 183.

14 Michael Hagner, 'Scientific Medicine', in David Cahan (ed.), *From Natural Philosophy to the Sciences: Writing the History of Nineteenth-Century Science* (Chicago: University of Chicago Press, 2003), 68.

15 'Succi: The Fasting Man', *The Lancet*, 10 Mar. 1888, 487.

16 'Nous connaissons ces effets par les expériences sur les animaux, et mieux encore par l'observation d'un certain nombre de jeûneurs 'professionnels'. Ces jeûneurs étaient en réalité à la diète hydrique, comme le sont les malades soumis à la cure de repos stomacal absolu, auxquelles on fournit aisément par la voie intestinales ou sous-cutanée l'eau qui leur est nécessaire; aussi leur étude est-elle, pour nous, particulièrement intéressante. Dès début de l'inanition, on observe chez les jeûneurs une forte perte de poids, qui va en s'atténuant peu à peu:

Chez Cetti ....600 gr. Par jour pendant 10 jours
Breithaupt........500 gr......................... 6 jours
Succi...............630 gr..................... les 10 premiers jours
" 390 gr..................... du 10e au 20e jour
" 290 gr..................... du 20e au 29e jour.'

See Georges Linossier, *Rapport sur le traitement de l'ulcère simple de l'estomac* (Paris: Masson, 1907), 6, www.biusante.parisdescartes.fr/histoire/biographies/?refbiogr=9978 (accessed 5 Jan. 2018).

17 '[R]ovinarsi la salute pel capriccio o per lo scopo di mostrare al pubblico sotto l'aspetto dell'uomo il più poetico del mondo è cosa che può destare qualche curiosità, far temere magari dell'avvenire ai venditori di commestibili, animare i dotti alle richerche, conciliare l'idea di un novello Prometeo che rapisca al cieto la favilla che lo anima, persuadere al pubblico dell'americanismo che penetra in Italia: è però un mètodo pericoloso di speculazione, una storia dolorosa ... ammazzarsi per vivere': Carlo Garampazzi and Giuseppe Raineri, *Il digiuno di G. Succi. Considerazioni fisico-patologiche* (Pallanza and Rome: Eredi Vercellini, 1886), 30; Giuseppe Raineri, 'Una visita al Succi', *Corrispondenza da Milano. Gazzette Piemontesca*, 5 Sep. 1886.

18 'Il Succi, con suo digiuno di 30 giorni, minaccia di sbalordire nos solo le persone ignoranti, quelle mediocramente o superlativamente istruite, ma perfino gli uomini della scienza togata': Tommaso Venanzi, *Il segreto del digiuno del Succi e del suo misterioso liquore spiegato al pubblico* (Rome: Domenico Vaselli, 1886), 5.
19 *British Medical Journal*, 21 Jun. 1890, 1446.
20 Neswald, 'Food Fights', 187.
21 Anton Julius Carlson, *The Control of Hunger in Health and Disease* (Chicago: University of Chicago Press, 1916), ch. IX, 'Hunger in Prolonged Starvation: Experiments on Men', 125. Dr Penny, who fasted for thirty days in 1909, drinking only distilled water, lost 19 per cent of his initial body weight of 137.5 pounds (62.4 kilograms): Francis Gano Benedict, *A Study of Prolonged Fasting* (Washington, D.C.: Carnegie Institution of Washington, 1915).
22 Hans-Jörg Rheinberger and Michael Hagner (eds.), *Die Experimentalisierung des Lebens. Experimentalsysteme in den Biologischen Wissenschaften, 1850–1950* (Berlin: Akademie Verlag, 1993).
23 Hans-Jörg Rheinberger, *Toward a History of Experimental Things: Synthesizing Proteins in a Test Tube* (Stanford: Stanford University Press, 1997); Lorraine Daston (ed.), *Things that Talk: Object Lessons from Art and Science* (New York: Zone Books, 2004).
24 Merriley Borell, 'Instruments and an Independent Physiology: The Harvard Physiological Laboratory, 1871–1906', in Gerald Geison (ed.), *Physiology in the American Context, 1850–1940* (Bethesda, Md.: American Physiological Society, 1987), 239–321.
25 William H. Brock, *Justus von Liebig: The Chemical Gatekeeper* (Cambridge: Cambridge University Press, 2002).
26 Soraya de Chadarevian, 'Graphical Method and Discipline: Self-Recording Instruments in Nineteenth-Century Physiology', *Studies in History and Philosophy of Science*, 24 (2) (1993), 267–291.
27 Geison (ed.), *Physiology in the American Context.* As the historiography of scientific instruments has rightly pointed out, the role of new devices usually changes when used for the study of living organisms. Sometimes instruments became an extension of the organism; in other cases, they acted as bridges between natural science and popular culture. In the case of experimental physiology, they conferred a scientific authority that doctors desperately sought. See Albert van Helden and Thomas L. Hankins, 'Introduction: Instruments and the History of Science', *Osiris*, 9 (1994), 5.
28 As stressed by Étienne-Jules Marey (1830–1904) in his *méthode graphique*, which appeared in 1878, just on the eve of the hunger artist period. See Étienne-Jules Marey, *La méthode graphique dans les sciences expérimentales et principalement en physiologie et en médecine ... 2e tirage augmenté d'un supplément sur le Développement de la méthode graphique par la photographie* (Paris: G. Masson, 1885); Merriley Borell, 'Instrumentation and the Rise of Modern Physiology', *Science and Technology Studies*, 5 (2) (1987), 53–62. Marey was a French physiologist and photographer. His contribution to chronophotographs became a source of inspiration for inventors such as Thomas Edison and the Lumière brothers.

29 P. Guarneri, 'Moritz Schiff (1823–1896): Experimental Physiology and Noble Sentiment in Florence', in Rupke (ed.), *Vivisection*, 105–124. For an English version of Mosso's work, see Angelo Mosso, *Fatigue* (London: Swan Sonnenschein & Co. Ltd, 1906). See also Anson Rabinach, *The Human Motor: Energy, Fatigue and the Origins of Modernity* (Berkeley: University of California Press, 1992).
30 'Ces messieurs commencèrent leurs études: l'un me sécha la bouche en m'enlevant la salive avec une petite pompe; un autre me brula les yeux en me comptant en jet lumineux dans chaque œil. Mais les plus curieux, ce furent les expériences de dynamométrie, consistant à me faire tirer sur une corde, à l'extrémité de laquelle n'était fixé un appareil en acier formant balance romaine': Monin and Maréchal, *Stefano Merlatti*, 122.
31 Dr Dutrieux-Bey: ibid., 211.
32 Ibid., 177, 209.
33 Ibid., 57.
34 Luciani, *Fisiologia del digiuno*, 28–118. Luciani referred to the canonical work of Dr Charles Chossat published in 1843, which established the loss rate of any tissue during inanition in animals.
35 Ibid., 44.
36 Ibid., 64–65.
37 Two years later, in 1890, during Succi's fast in London, the dynamometer measured his strength (49 kg at the beginning of his fast and 50 kg on the fourth or fifth day), his breathing capacity (1500–1600 cc at the beginning and 1500 cc on the forty-fifth day), his weight loss (35 pounds in a forty-day fast), and his regular and firm pulse, all while smoking an average of twenty cigarettes per day. Instruments provided data on those apparently inexplicable records, which made the case relevant to many doctors, and presented an objective, reliable image of the quantified data. See 'Doctors Wonder at Succi's Mystery', *New York Herald*, 8 Nov. 1890. The outcome of his fast was also described in the following terms: 'When Succi began his fast he weighed 147 1/4 pounds and was five feet and four inches tall. After living 30 days on nothing but air and water the man was reduced in weight to 112 1/4 pounds and his height had decreased one half of an inch. He has lost 21 1/2 inches around the abdomen and 10 inches on the thighs. The animal fat he accumulated previous to beginning the fast is gone, and the question to be determined is whether Succi has enough muscular tissue on his frame to keep himself alive the remaining time': 'A Case of Starvation: Signor Giovanni Succi, Who Is Living without Food', *Springfield Republican*, 7 Dec. 1890.
38 'With the new model of von Basch's blood pressure monitor, in which the mercury manometer is suitably replaced by a highly sensitive aneroid manometer, I have regularly explored the blood pressure in the right radial, morning and evening, on each of the thirty days of fasting ... [The diagram] invites us to consider two facts: first, a slow and gradual decrease in blood pressure from the first to the last days of fasting, is lowered to a minimum of 12 cm; second, some daily fluctuations from morning to evening, which often, but not always, correspond to homonymous fluctuations of the temperature and heart rate curves': Luciani, *Fisiologia del digiuno*, 32.

39 Ibid., 19–141.

40 '[C]onsta dei seguenti pezzi:

a) Una specie d'imbuto di gomma elastica adatta allá forma delle arcate dentali, e che s'introduce tra queste e la labbra, le quali fanno da otturatori.
b) Al centro di detto imbuto è annesso un largo tubo di vetro che si biforca in due, congiunti questi mercè tubi flessibili di gomma, con due apparecchi valvolari del Müller, colla interpolazione di due tubi di vetro a T chiusi nella branca inferiore, che raccolgono la saliva che potesse eventualmente scolare dalla bocca.
c) Le due valvole offrono pochissime resistenze e sono disposte in direzione inversa, di guisa che una si apre nell'inspirazione e si chiude nell'espirazione, l'altra viceversa si apre nell'espirazione e si chiude nell'inspirazione.
d) La valvola *inspiratoria* contiene una soluzione di potassa caustica al 50% et è congiunta con due alti cilindri ripieni di pezzi di pomice imbevuta di una soluzione di potassa, e di qualche pezzeto di potassa solida. Questi cilindri allá lor volta sono congiunti con un contatore molto sensibile, che segna la quantità di aria inspirata.
e) La valvola *espiratoria* contiene acqua distillata, ed è congiunta con un sacco di tela gommata della capacità di 100 litri, destinato a raccogliere tutta l'aria, satura di umidità, *espirata* in 10 minuti di respirazione nell'apparechio.
f) Un pneumografo del Marey applicato al torace del digiunatore traccia in un cilindro girante il numero e le escursioni degli atti respiratori che si compiono nei 10 minuti di durata de ciascuna ricerca.'

See Luciani, *Fisiologia del digiuno*, 123.

41 Frederick L. Holmes, 'The Intake–Output Method of Quantification in Physiology', *Historical Studies in the Physical and Biological Sciences*, 17 (2) (1987), 247.

42 Luigi Luciani, *Das Hungern. Studien und Experimente am Menschen* (Hamburg and Leipzig: Leopold Voss, 1890).

43 Nieto-Galan, 'Hunger Artists and Experimental Physiology'.

44 Luciani, *Fisiologia del digiuno*, 147–148.

45 Luciani, *Human Physiology*, 421; Ermanno Manni and Laura Petrosini, 'Luciani's Work on the Cerebellum, a Century Later', *Trends in Neurosciences*, 20 (1997), 112–116; Carmela Morabito, 'Luigi Luciani and the Localization of Brain Functions: Italian Research within the Context of European Neurophysiology at the End of the Nineteenth Century', *Journal of the History of the Neurosciences*, 9 (2000), 180–200. This was a general pattern that Luciani developed later and published some years afterwards: Luigi Luciani and Silvestre Baglioni, *L'alimentazione umana secondo le piú recenti indagini fisiologiche* (Milan: Società Editrice Libraria, 1917).

46 *The Lancet* described Luciani's experiments on Succi in Florence in the following terms: 'The result is watched with special interest by Italian surgeons and physicians anxious to get some scientific evidence of the time during which the cyclo-poietic viscera can remain undisturbed by the ingestion of

food – a point of the highest importance in abdominal surgery or in bowel haemorrhage from perforation' ('Succi, the Fasting Man', 487).

47 For the early experimental culture of nutrition, see Carl Ludwig, *Lehrbuch der Physiologie des Menschen* (Heidelberg: Winter, 1852).

48 Pettenkofer, Rubner, Zuntz, Sondén, Tigerstedt, Atwater, Rosa, and Benedict: W. O. Atwater and F. G. Benedict, *A Respiratory Calorimeter with Appliances for the Direct Determination of Oxygen* (Washington: Carnegie Institution of Washington, 1905), Publication No. 42. At the Munich Physiological Institute, Pettenkofer built: 'a new apparatus to measure respiration on an unprecedented scale, yet with greater precision than any apparatus that had previously existed' (Holmes, 'The Intake–Output Method of Quantification in Physiology', 261). For the repercussions of the respiration calorimeter in Mexico, see Joel Vargas, 'Calibrar la alimentación. La estandarización del calorímetro en México', in Laura Cházaro, Miruna Achim, and Nuria Valverde (eds.), *Piedra, papel y tijera. Instrumentos en las ciencias en México* (Mexico City: UAM Cuajimalpa, 2018), 211–244.

49 Max von Pettenkofer (1818–1901) was a chemist and hygienist. A pupil of Justus von Liebig in Giessen, he was later appointed chair of Pathological Chemistry in Munich. He also contributed to the design of the respiration calorimeter to study metabolism in humans. See Mathias Schütz, 'After Pettenkofer: Munich's Institute of Hygiene and the Long Shadow of National Socialism, 1894–1974', *International Journal of Medical Microbiology*, 310 (5) (2020), 1–7.

50 Lara Marks, *Useful Bodies: Humans in the Service of Medical Science in the Twentieth Century* (Baltimore: Johns Hopkins University Press, 2003). 'The air was aspirated through the chamber, and at the point of exit samples were measured, after having been passed through suitable solutions for the removal of carbon dioxide and water': Mary Swartz Rose, *The Foundations of Nutrition* (New York: Macmillan Company, 1927), 14, 15.

51 'Energy values to the body of starch and fat were equal to the heat produced by burning them in a special apparatus for heat determinations, called a calorimeter; but the energy value of protein was different, since it could not be as completely burned in the body as it could in the calorimeter': Rose, *The Foundations of Nutrition*, 14, 15. See C. A. Galbraith, 'Wilbur Olin Atwater', *Journal of Nutrition*, 124 (9) (1994), 1715–1717. See also Elizabeth Neswald and David F. Smith, *Setting Nutritional Standards: Theory, Policies, Practices* (Rochester, N.Y.: Rochester Studies in Medical History, 2017).

52 Francis Gano Benedict, *The Influence of Inanition on Metabolism* (Washington, D.C.: Carnegie Institution of Washington, 1907), 7–10. See also Francis Gano Benedict, 'The Nutritive Requirements of the Body', *American Journal of Physiology*, 16 (1906), 409. Benedict referred to earlier experiments on the exchange of nitrogen, carbon dioxide, water, and heat in a respiration calorimeter, with his master Atwater. He mainly relied on a group of volunteer medical students, and professional faster referred to as SAB. 'The subjects, were, with the exception of SAB, all students in Wesleyan University, nine young men [BFD, ALL, HES, CRY, AHM, HCK, HRD, NMP, DW], in good Health. SAB, with whom the longer experiments were made, was a young masseur, who had made several fasts prior to his arrival in Middletown

with a view of studying his daily losses of weight, who spent 24 hours of each day inside the respiration chamber': Benedict, *The Influence of Inanition on Metabolism*, 10; Benedict, *A Study of Prolonged Fasting*, 26. Benedict classified his experiments into gross observations and physiological, chemical and physical approaches. General observations included body weight measurements and an examination by a physician measuring blood pressure and pulse with a sphygmomanometer. Physiological measurements included body temperature with a rectal thermometer (bolometer) and pulse rate. This was followed by the chemical analysis of food, faeces (water, ash, nitrogen, carbon, hydrogen, ether, sulphur, and phosphorus), urine (chlorine, phosphoric acid, sulphuric acid, creatine, and creatinine), and respiration products (water, carbon dioxide, and oxygen). Physical measurements consisted of the potential energy of food, faeces, and urine with a bomb calorimeter, heat elimination, and also the use of the respiratory calorimeter. See Rose, *The Foundations of Nutrition*, which includes a historical chapter and an impressive collection of images.

53 'Mr T. M. Carpenter, in immediate supervision of the greater number of the respiration calorimeter experiments, has conducted these most wearisome experiments with unusual success. Mr H. A. Pratt, aside from rendering valuable assistance in the chemical laboratory, has superintended the computations and tabulations, and the entire report has received his helpful editorial criticism. Mr E. M. Swett acted as physical and chemical assistant and superintended the determinations of the heats of combustion with the bomb calorimeter. Miss Charlotte E. Manning had charge of all the gas analyses, carbon and hydrogen combustions, creatinine determinations, and the analyses of food. Mr E. E. Fulton made all the determinations of sulphur and phosphorus. Mr F. P. Fletcher acted as physical assistant in the later series of calorimeter experiments. Messrs J. A. Eiche and E. E. Ilartman assisted in the chemical and physical measurements. Mr W. H. Leslie, Miss A. N. Darling, and Mr H. C. Morgan have had a large share in the tabulation of the results of the experiments and in the final preparation of the report. The stenographic work was in charge of Miss A. N. Darling, who was ably assisted by Miss M. K. Falsey. Aside from a corps of students, special mention should be made of the assistance in the computations rendered by Messrs H. L. Knight and F. W. Harder and Misses H. W. Atwater, E. J. Wright, and H. L. Ailing': Benedict, *The Influence of Inanition on Metabolism*, 1.

54 Neswald, 'Food Fights', 178.

55 Ibid., 179–180. 'These experimental subjects were both conscious, intentionally acting agents and bodies that neither the subject nor the experimenter could fully control. Bodies as well as subjects cooperated and resisted, and both needed to be accommodated for an experiment to succeed': ibid., 176.

56 *Virchows Archiv für pathologische Anatomie und Physiologie und für klinische Medizin*, 131 (1893), p. 1.

57 Monin and Maréchal, *Stefano Merlatti*.

58 'Réunis dans l'appartement qu'il occupe au première étage du Grand Hôtel, à Paris, et qui a été mis gracieusement à disposition par la direction du dit hôtel, en présence de MM. Mario Carl-Rosa, directeur de l'Académie des

beaux-arts des Champs Elysées, et Ragot, artiste peintre; après avoir pris connaissance du journal de l'expérience et des analyses d'urine faites par M. Edmond Vasseur, pharmacien … on décide que, malgré le bon état relatif du sieur Stefano Merlatti, il y aurait peut-être quelque danger à continuer, dans un but de simple curiosité, un jeûne absolu qui a déjà duré, sans aucune interruption, huit jours': ibid., 172.

59 'On se souvient du jeûneur Merlatti. Le malheureux est décédé à la suite gastrite contractée pendant ses jeûnes. Il aimait trop le jeûne; c'est ce qui l'a tué. Si son sort servait, au moins, d'exemple à Succi': 'Mort de Merlatti', *L'Avenir de Bel-Abbès. Journal agricole, commercial, industriel, politique et littéraire*, 3 May 1890, 1.

60 Equally, in his treatise on physiology, Charles Richet also discussed the risks Merlatti took during his fast to the limits of human resistance after disregarding his medical committee's advice: 'malgré de jeûne prolongé, [Merlatti] résistait au conseil qu'on lui donnait d'arrêter son expérience' (Richet, *Physiologie. Travaux du laboratoire. Chimie physiologique. Toxicologie* (Paris: Félix Alcan, 1893–1895), ch. 29, 'L'inanition', 305.

61 Chossat, *Recherches expérimentales sur l'inanition.*

62 Austin Flint in New York, for example, quoted Chossat when discussing the frontier between fasting, starvation, and death in the following terms: 'Death from starvation occurs after a loss of 4/10 of the weight of the body, the time of death being variable in different classes of animals' (Austin Flint, *A Text-book of Human Physiology*, 4th ed. (London: H. K. Lewis, 1888), 169).

63 Luciani, *Fisiologia del digiuno*, 8.

64 Claudio Pogliano, 'La fisiologia in Italia fra ottocento e novecento', *Nuncius*, 6 (1) (1991), 115–116; Nieto-Galan, 'Hunger Artists and Experimental Physiology'.

65 '[LUCIANI] – Apprezzo l'originalità e simplicità dei mezzi da voi impiegati per far fortuna. Coi vostri digiuni voi siete riescito a conquistarvi una rinomanza …

[SUCCI] – Ma non ancora a formarmi una fortuna, caro signor Professore.

[…]

[LUCIANI] – Che cosa ne vorreste concludere? Tenendo conto soltanto della vostra costituzione più robusta, io mi spiego benissimo la vostra maggior resistenza rispetto al Merlatti. Ma perchè voi non metteste in maggior evidenza la vostra superiorità prolungando il digiuno oltre i termini fissati dal Merlatti?

[SUCCI] – Lo vuol sapere perchè? Per dimostrare che ho quattro dita di cervello e che non sono niente affato un matto. Io non volli consumare la mia carne per un semplice puntiglio. Il mio contratto era di diuginare per 30 giorni; scaduto il termine fissato, io mi tenni sciolto da qualsiasi obbligo.'

See Luciani, *Fisiologia del digiuno*, 25.

66 I discuss the Florence experiments in detail in Nieto-Galan, 'Hunger Artists and Experimental Physiology'.

67 *Yorkshire Post and Leeds Intelligencer*, 14 Mar. 1887, 5.

68 Curt Lehmann, Friedrich Müller, Immanuel Munk, Hermann Senator, and Nathan Zuntz, 'Untersuchungen an zwei hungernden Menschen', *Archiv für pathologische Anatomie und Physiologie und für klinische Medizin*, 131, supplemente Heft, 1893; 'Professor Virchow über Cetti', *Berliner Zeitung*, 27 Mar. 1887; Nina Diezemann, *Die Kunst des Hungerns. Essstörungen in Literatur und Medizin um 1900* (Berlin: Kadmos, 2006), 72–80.
69 Lehmann, et al., 'Untersuchungen an zwei hungernden Menschen'.
70 Ibid. (Conclusions). See also Graham Lusk, *Elements of the Science of Nutrition* (Philadelphia: W. B. Saunders, Co., 1906), ch. 2, 'On Starvation', 62–63.
71 Lehmann, et al., 'Untersuchungen an zwei hungernden Menschen', 'Schlusswort'.
72 Benedict, *A Study of Prolonged Fasting*, 20.
73 Elizabeth Neswald, 'Strategies of International Community-Building in Early Twentieth-Century Metabolism Research: The Foreign Laboratory Visits of Francis Gano Benedict', *Historical Studies in the Natural Sciences*, 43 (1) (2013), 1–40.
74 Benedict, *The Influence of Inanition on Metabolism*, 1.
75 Levanzin had previously contacted Luciani, who advised him to go to Boston and contact Benedict, who finally brought Levanzin to his laboratory.
76 Benedict, *A Study of Prolonged Fasting*, 28; Sharman Apt Russell, *Hunger: An Unnatural History* (New York: Basic Books, 2005), 57.
77 Benedict, *A Study of Prolonged Fasting*, 213. 'Esperanto had a great vogue in Malta; I, with Mrs Levanzin, took part in the International Congress of Barcelona and there I was elected "President of the International Association of Pharmaceutical Esperantists", editor of the scientific Esperanto monthly, "La Vocho de Farmacustoj", and Corresponding Member of the "Colegio des Farmaceuticos"': https://timesofmalta.com/articles/view/Agostino-Levanzin-man-for-all-reasons.428703 (accessed 10 Apr. 2020).
78 Benedict, *A Study of Prolonged Fasting*, 378 (see also all the final tables on Levanzin).
79 Lederer, *Subjected to Science*, 118–119. For an evaluation of the tensions between Benedict and Levanzin in the press, see 'Torture in Starving Test: Levanzin Complains of Treatment at Carnegie Nutrition Laboratory', *New York Times*, 21 May 1912. See also *New York Times*, 5 May 1912, 11; 15 May 1912, 24; 22 May 1912, 11; 27 May 1912, 4; Benedict to Robert Woodward, 3 May 1915, Carnegie Institution of Washington (CIW) Archive, Nutrition Laboratory Box 1, File 15; Agostino Levanzin to Woodward, 12 Jun. 1912, CIW Archive, Box 1, File 15; Woodward to Benedict, 14 Jun. 1912, CIW Archive, Nutrition Laboratory, Box 1, File 14. See also Neswald, 'Food Fights'.
80 *New York Times*, 25 May 1912.
81 Francis Gano Benedict, 'Chemical and Physiological Studies of a Man Fasting Thirty-One Days', *Proceedings of the National Academy of Science of the United States of America*, 1 (1915), 228–231.
82 'Confined in Coffin: Man Held for Experiments at the Carnegie Institution; Harvard Student Gives Details of Professor Levanzin's Experiments; Man Placed in Air-Tight Box for 33 Days without Food', *Lyons Republican*, 18 Jun. 1912, 3.
83 *Mount Union Times*, 5 Jul. 1912, 7.

84 'Augustin Levanzin of Los Angeles was fined $500 and given a suspended sentence of 80 days in jail, in a charge of practicing medicine without a license. Levanzin, a native of Malta, is a scholar and author, holding several European diplomas': 'Medical Persecution', *Los Angeles Times*, 15 Aug. 1920, 160.
85 Benedict, *A Study of Prolonged Fasting*, 11.
86 Ibid. (my emphasis).
87 'Today is of great interest to elucidate the underlying mechanisms enabling metabolic adaptations occurring during prolonged fasting and refeeding': Jean-Hervé Lignot and Yvon LeMaho, 'A History of Modern Research into Fasting, Starvation and Inanition', in Marshall D. McCue (ed.), *Comparative Physiology of Fasting, Starvation, and Food Limitation* (Dordrecht: Springer, 2012), 16.
88 Sergius Morgulis, *Fasting and Undernutrition: A Biological and Sociological Study of Inanition* (New York: Dutton and Company, 1923), 3.
89 Benedict, *The Influence of Inanition on Metabolism*, 2 (my emphasis).
90 Benedict, 'Chemical and Physiological Studies', 228.
91 When measuring the output of total nitrogen in the metabolic process, starvation experiments used animals and humans, and in the latter case many were professional hunger artists: Edward P. Cathcart, *The Physiology of Protein Metabolism* (London: Longmans, Green and Co., 1912), 96.
92 David Cahan, *Helmholtz: A Life in Science* (Chicago: University of Chicago Press, 2018).
93 Luciani, *Das Hungern*, 1890; Lo Monaco, *Bullettino della Società Lancisiana degli ospedali di Roma*, 14 (1894), 102; Daiber, 'Beitrag zur Kenntnis des Stoffwechsels beim Hungern', *Schweizer Wochenblatt für Chemie und Pharmacie*, 34 (1896), 395; E. Freund and O. Freund, 'Beitrage zum Stoffwechsel im Hungerzustande', *Wiener klinische Rundschau*, 5–6 (1901); D. Baldi, *Zentralblatt für klinische Medizin*, 10 (1889), 651; Ajello and Solano, *La riforma medica*, 1 (2) (1893), 542; Tauszk, *Orovosi hetilap*, 512 (1894); and finally Brugsch, 'Über die Rolle des Glykokolls im intermediaren Eiweissstoffwechsel beim Menschen', *Zentralblatt für die gesamte Physiologie und Pathologie des Stoffwechsels*, 8 (1907), 529; 'Eiweisszerfall und Acidosis im extremen Hunger mit besonderer Berücksichtigung der Stickstoffverteilung im Harn', *Zeitschrift für experimentale Pathologie und Therapie*, 1 (1905), 419; Brugsch and Hirsch, 'Gesammtstickstoff- und Aminosdurenausscheidung im Hunger', *Zeitschrift für experimentale Pathologie und Therapie*, 3 (1906), 3, 638.
94 Lehmann, et al., 'Untersuchungen an zwei hungernden Menschen'; Hermann Senator, 'Bericht über die Ergebnisse des an Cetti augeführten Hunger versucher. Verhalten der Organe und swa Atoffwechsels im allgemeinen', *Biologische Zentralblatt*, 7 (1887), 344–352.
95 Langworthy was at that time the chief of the Bureau of Home Economics of the US Department of Agriculture.
96 W. O. Atwater and C. F. Langworthy, *A Digest of Metabolism Experiments in Which the Balance of Income and Outgo Was Determined* (Washington, D.C.: US Government Printing Office, 1898), 88–91.
97 Ibid., 88–92.

98 Morgulis, *Fasting and Undernutrition*, 163; Cathcart, *The Physiology of Protein Metabolism*; Edward P. Cathcart, 'Ueber die Zusammensetzung des Hungerharns', *Biochemische Zeitschrift*, 6 (1907).
99 Cathcart, *The Physiology of Protein Metabolism*, 96, 120.
100 Eugene F. Du Bois, 'Biographical Memoir of Graham Lusk', *National Academy of Sciences of the United States of America: Biographical Memoirs*, vol. XXI, 1939, 93–142.
101 Lusk, *Elements of the Science of Nutrition*.
102 Ibid., 62–64.
103 Ibid., 80.
104 Lusk, *Elements of the Science of Nutrition*, 64.
105 Benedict, 'Chemical and Physiological Studies', 230.
106 Ibid., 229.
107 Morgulis, *Fasting and Undernutrition*; Cathcart, 'Ueber die Zusammensetzung des Hungerharns'. Lusk cited von Voit in the following terms: 'Death from starvation is primarily due to loss of substance in organs important to life, but it may also ensue under certain circumstances, as a result of deficient nutrition of these organs' (Lusk, *Elements of the Science of Nutrition*, 75).
108 Morgulis, *Fasting and Undernutrition*. Morgulis dedicated the book to Thomas Hunt Morgan (1866–1945), the future Nobel laureate in Physiology or Medicine in 1933 for discovering the role of chromosomes. The introduction to the book first appeared in the *Scientific Monthly*.
109 Lignot and LeMaho, 'A History of Modern Research into Fasting', 10. Some years earlier, in 1890, Charles Richet had already described four analogous stages in the inanition process: first, the consumption of the ingested food; secondly, the use of carbohydrates and fat reserves; thirdly, the consumption of albuminoid tissues, which can be measured by an increase of nitrogen in urine; and, finally, the premortal phase. See Henri Labbé, 'L'inanition. Ses aspects physiologiques et sociaux', *Revue scientifique*, 2 May 1890, 549.
110 Morgulis, *Fasting and Undernutrition*, viii.
111 Ibid., 163 (comparative tables).
112 Lignot and LeMaho, 'A History of Modern Research into Fasting', 7–8.
113 'The Fasting Man: the objects of the fast; Mr Robin's report; the clinical events; absence of all serious symptoms; Succi's elixir; his beverages; his condition at the end of the fast; his subsequent progress; report of Messrs Powell, Burt and Pike; the state of the secretions; the pulse and blood; mental agitation and loss of body weight; the urine and its constituent; the body temperature; concluding observations: cui bono?', *British Medical Journal*, 21 (1890), 1446.
114 Morgulis, *Fasting and Undernutrition*, 9.

4 Spirits

1 Michael Hagner, 'Scientific Medicine', in David Cahan (ed.), *From Natural Philosophy to the Sciences: Writing the History of Nineteenth-Century Science* (Chicago: University of Chicago Press, 2003), 50.
2 Ibid., 65.

3 John Warne Monroe, *Laboratories of Faith: Mesmerism, Spiritism and Occultism in Modern France* (Ithaca: Cornell University Press, 2008), 12. See also Richard Noakes, *Physics and Psychics: The Occult and the Sciences in Modern Britain* (Cambridge: Cambridge University Press, 2019).
4 Bernadette Bensaude-Vincent and Christine Blondel (eds.), *Des savants face à l'occulte, 1870–1940* (Paris: La Découverte, 2002); Lisa Abend, 'Specters of the Secular: Spiritism in Nineteenth-Century Spain', *European History Quarterly*, 31 (4) (2004), 507–534.
5 Bensaude-Vincent and Blondel (eds.), *Des savants face à l'occulte.*
6 'Dans de siècle de positivisme on a nié tout surnaturel, le rationalisme a voulu tout expliquer par le naturalisme, et voilà que jamais, depuis le Moyen Age, le surnaturel ne s'est jamais tant affirmé sous tous les yeux, à toutes les portes': L'abbé Vallée, curé de Monts, 'Essai d'explication théorique et scientifique du jeûne extraordinaire des italiens Succi et Merlatti', *Annales de la Société d'agriculture, sciences, arts et belles lettres d'Indre-et-Loire*, 67 (1887), 35.
7 Bensaude-Vincent and Blondel (eds.), *Des savants face à l'occulte.*
8 George Weisz, 'The Emergence of Medical Specialization in the Nineteenth Century', *Bulletin of the History of Medicine*, 77 (2003), 536–575.
9 Régine Plas, 'Psychology and Psychical Research in France around the End of the Nineteenth Century', *History of the Human Sciences*, 25 (2) (2012), 91–107.
10 Kurt Dazinger, 'Wilhelm Wundt and the Emergence of Experimental Psychology', in Robert Olby, et al. (eds.), *Companion to the History of Modern Science* (London: Routledge, 1990), 396–409.
11 'Some time ago Dr Henry S. Tanner was arrested in Jamestown, NY, for practising Medicine in violation of the law of the State': *Medical Tribune*, 15 Jan. 1884, 36.
12 Robert A. Gunn, *40 Days without Food: A Biography of Henry S. Tanner MD, Including a Complete and Accurate History of His Wonderful Fast, 42 Days in Minneapolis, and 40 Days in New York City, with Valuable Deductions* (New York: Albert Metz, 1880), 8.
13 Sharman Apt Russell, *Hunger: An Unnatural History* (New York: Basic Books, 2005), 54–55.
14 Marij van Strien, 'Vital Instability: Life and Free Will in Physics and Physiology, 1860–1880', *Annals of Science*, 72 (3) (2015), 381–400; Frederick Gregory, *Scientific Materialism in Nineteenth-Century Germany* (Dordrecht: Riedel, 1977).
15 Van Strien, 'Vital Instability', 384.
16 Ibid., 388; Gabriel Finkenstein, *Émile du Bois-Reymond: Neuroscience, Self, and Society in Nineteenth-Century Germany* (Cambridge, Mass.: MIT Press, 2013).
17 Elizabeth A. Williams, *A Cultural History of Medical Vitalism in Enlightenment Montpellier* (London: Routledge, 2016).
18 Discovered by Claude Bernard, glycogen is a multi-branched polysaccharide of glucose that serves as a form of energy storage in animals. It is 'quickly used up with the onset of inanition, only traces remaining in the tissues, [and] the organism subjected to protracted starvation becomes an essentially fat–protein system': Sergius Morgulis, *Fasting and Undernutrition: A Biological and Sociological Study of Inanition* (New York: Dutton and Company, 1923), 127.

19 Williams, *A Cultural History of Medical Vitalism.*

20 Claudio Pogliano, 'La fisiologia in Italia fra ottocento e novecento', *Nuncius*, 6 (1) (1991), 115–118.

21 Ibid., 99.

22 'La regolazione della nutrizione e della termogenesi, dei processi d'integrazione e di disintegrazione, o più generalmente, dello scambio materiale e dinamico, sia di ciascuna parte che del complesso dell'organismo, è funzione fondamentale del sistema nervoso, considerato nel suo insieme e nella sua unità, é non di una o d'altra parte o segmento di esso sistema': Luigi Luciani, *Fisiologia del digiuno: studi sull'uomo* (Florence: Tipografia dei Successori Le Monnier, 1889), 157.

23 Elizabeth A. Williams, *Appetite and Its Discontents: Science, Medicine and the Urge to Eat, 1750–1950* (Chicago: University of Chicago Press, 2020).

24 Agustí Nieto-Galan, 'Useful Charlatans: The Fasting Contest of Giovanni Succi and Stefano Merlatti in Paris, 1886', *Science in Context*, 33 (4) (2020), 405–422.

25 Nicole Edelman, *Histoire de la voyance et du paranormal. Du XVIIIe siècle à nos jours* (Paris: Seuil, 2006); Sofie Lachapelle, *Investigating the Supernatural: From Spiritism and Occultism to Psychical Research and Metapsychics in France, 1853–1931* (Baltimore: Johns Hopkins University Press, 2011). I am indebted to Mònica Balltondre and Andrea Graus for these references.

26 Annette Mulberger and Mònica Balltondre, 'Metapsychics in Spain: Acknowledging or Questioning the Marvellous?', *History of Human Sciences*, 25 (4) (2012), 109; Lachapelle, *Investigating the Supernatural.*

27 Lynn L. Sharp, *Secular Spirituality: Reincarnation and Spiritism in Nineteenth-Century France* (Plymouth: Lexington Books, 2006), xix–xx.

28 Allan Kardec, *Le livre des esprits. Contenant les principes de la doctrine spirite sur l'immortalité de l'âme, la nature des esprits et leurs rapports avec les hommes … recueillis et mis en ordre par Allan Kardec*, 15th ed. (Paris: Didier, Dentu, 1867 [1857]).

29 'The lecture, which was benevolently commented upon everywhere, and illustrated by the experiments of auto-suggestion, of hypnotism, was carried out by the young professor William Etrusco who has been following Succi for some time': *Cenni Biografici. Giovanni Succi. Esploratore d'Africa. Già delegato della 'Società di Commercio coll'Africa' di Milano e membro della 'Società psicologica di Madrid'* (Rome: Tipografia Económica, 1897), 13.

30 After drinking a mixture of laudanum and morphine – two ingredients of Succi's liquor.

31 Luciani, *Fisiologia del digiuno*, 17–19.

32 '[N]elle loro notturne sedute ricevevano comunicazioni che un forte, detto *Spirito Leone*, era arrivato a Roma … iniziali G.S.': ibid., 17.

33 'Succi, tu es celui qui peut vivre sans se nourrir … tu peux défier chevaux et chevaliers quand mon rayon lumineux se reflète sur toi … Dans toutes les langues qui s'impriment, ton nom sera lu': *L'Écho du merveilleux*, 1 Oct. 1902, 380.

34 'Le voyageur avait eu à Zanzibar un nègre dévoué qui le sauva d'une maladie mortelle avec certaines herbes. Ce fidèle serviteur fut tué dans une rixe. Succi désolé revint en Italie avec ses herbes. Durant une séance de spiritisme à

Rome, il eût l'idée d'évoquer l'esprit de son regretté nègre et celui-ci, qui sans doute, n'avait pas perdu les habitudes de respectueuse obéissance, répondit à l'appel. Divers amis de Succi, qui assistaient à cette séance, affirment que le voyageur en sortant était dans un état de surexcitation inquiétante. L'Esprit lui avait conseillé de tenir grand comte des herbes recollées à Zanzibar, parce qu'elles lui serviraient un jour à faire une découverte qui révolutionnerait l'humanité en supprimant la faim': *La vie posthume*, 8 (1888), 192.

35 '[L]a sua fede spiritistica sulla quale fondava la sua resistenza ai lunghi digiuni e la sua pretesa refrattarietà ai veleni (di cui però non volle fornirci la prova) sono senza dubbio un sedimento residuale del suo delirio, e non ci permettono di escludere il pericolo di un futuro risorgere de' suoi sogni ambiziosi': Luciani, *Fisiologia del digiuno*, 27.

36 '[LUCIANI] – E sarebbe? ... Lo spiritismo? ... Via, caro Succi, parliamoci in cofidenza ... Vi sembre serio che un uomo come voi, che ha tanto viaggiato e che mostra tanto giudizio, si perda in certe ubbie?

[SUCCI] – Io non ho la pretesa di convertirla allo spiritismo ... anzi se vorrà lasciar da parte questo argomento mi farà piacere.

[LUCIANI] – Ma scusate, il Merlatti che digiunò per 40 giorni contemporaneamente a voi, era un spiritista anche lui?

[SUCCI] – E per questo ci fece una bella figura! Non sa Lei che a un certo periodo fu abbnadonato dai medici, che non vollero assumere la responsabilità della continuazione del digiuno? Agli ultimi giorni era ridotto a tali estremi, che per tenerlo ben caldo dovettero involgerlo di bambagia! Quello no fu un esperimento; fu un vero temptativo di suicidio che le autorità avrebbero dovuto impedire. [...]

[LUCIANI] – Non vi posso dar torto. Ma intanto voi pure dovete convenire che si può digiunare non solo per 30, ma anche per 40 giorni, senza l'aiuto di nessuno *spirito*, nemmeno dello *spirito di-vino.*

[SUCCI] – Lei vuol scherzare ... La prego, non torniamo su quest'argomento.'

See Luciani, *Fisiologia del digiuno*, 25–26.

37 'Giovanni Succi dà ad intendere di credere nello Spiritismo, ma proprio in quello il più delirante; e [...] questa è una delle grullerie che Egli adopera per fare effetto nel volgo, senza punto affatto credervi': Angiolo Filippi, 'Il Sor Giovanni Succi digiunatore e l'Academia Medico Fisica fiorentina', *Lo Sperimentale*, vol. 61, Mar. 1888.

38 *Il Corriere Spiritico. Rivista Mensile Scientifica dello Spiritismo. Direttore: Giovanni Succi; Gerente Responsabile: Achile Ricci. Direzzione e Redazione della rivista: Piazza della Signoria, 7. Firenze* (issues 1–6, 1888, are preserved at the Biblioteca Marucelliana in Florence). In May 1888, a new title appeared: *Il Corriere Spiritico. Giornale Scientifico dello Spiritismo e Magnetismo. Rivista Quindicinale. Fondata e diretta de Giovanni Succi* (Florence: Tipografia Coppini e Bocconi, 1888). For more details, see Nieto-Galan, 'Hunger Artists and Experimental Physiology in the Late Nineteenth Century: Mr Giovanni Succi Meets Dr Luigi Luciani in Florence', *Social History of Medicine*, 28 (1) (2015), 82–107.

39 'Una forza omnipotente, che i materialisti negano ...': 'Giovanni Succi', *Il Corriere Spiritico*, 4 (1888): 103–104.

40 'E quando il nostro Direttore Giovanni Succi avrà qui in Firenze intrapeso e compiuto il suo digiuno, noi faremo conoscere quale è il metodo e l'applicazione della forza spiritica per cui si puo raggiungere questo grandioso intento, uno dei ritrovati più strani e di induscutible utilità, destinato ad apportare *una rivoluzione nella scienza fisiologica e psicologica attuale e nel benesse sociale*': *Il Corriere Spiritico*, 1 (1888), p. 2 (my emphasis).
41 'Che cosa è la vita? Per i materialist è un sogno che passa e più non ritorna, ma per chi sa che cosa sia l'Io, la vita è il reale perchè l'Io è immortale'; 'Che cosa è la morte? Sublime Lancia che ferma chi troppo si innalza'; 'non è uomo, ossia essere umano, la materia che copre l'uomo, il vero essere è quello che sta rinchiuso entro la Maschera dell'uomo': *Il Corriere Spiritico*, 4 (1888), 105–107.
42 Thomas P. Weber, 'Carl du Prel (1839–1899): Explorer of Dreams, the Soul, and the Cosmos', *Studies in History and Philosophy of Science Part A*, 38 (3) (2007), 593–604.
43 Turiello, *Saggio dello Spiritismo in Italia* (Naples: Giuseppe Golia, 1898).
44 Ibid.
45 'Il sopranaturale non existe. La manifestazione attenuate con il mezzo dei medium, comme quelle del magnetismo e del sonambulismo, sono nell'ordine naturale delle cose e devono essere sottomesse severamente al crontrollo dell'sperienze. Non vi sono più miracoli ... [ma] lo studio positive di questa nuova psicologia': *Il Corriere Spiritico*, 1 (1888), p. 1.
46 Paul Gibier, *Le spiritisme (fakirisme occidental). Étude historique, critique et expérimentale (4e édition revue et corrigée) par le Dr Paul Gibier, Ancien interne des hopitaux de Paris, Aide naturaliste au Muséum d'histoire naturelle* (Paris: Octave Doin, 1896; 1st ed., 1887), 181–184.
47 For details on the role of science and technology in the 1888 Barcelona International Exhibition, see Agustí Nieto-Galan, 'Scientific "Marvels" in the Public Sphere: Barcelona and Its 1888 International Exhibition', *HoST – Journal of History of Science and Technology*, 6 (2012), 7–38.
48 *Cenni Biografici. Giovanni Succi*, 8.
49 'Léese una comunicación del Sr Succi, pidiendo al Congreso nombre una comisión para comprobar y fiscalitzar su ayuno de treinta días y los ejercicios corporales': *La Exposición*, 25 Sep. 1888, 7.
50 *El Criterio Espiritista. Revista mensual, órgano oficial de la Sociedad Espiritista Española. Fundador: Alverico Perón* (Madrid: T. Fontanet, 1868–1878).
51 'Esperemos pues que no encontrará la humanitaria y regeneradora doctrina del espiritismo insoportables trabas oficiales, y que sus adeptos no serán perseguidos por dedicarse a su propagación. Tropezarán sin duda todavía con el obstáculo de la incredulidad, pero esta oposición es poco temible, porque de día en día se debilita, y el materialismo no tiene raíces en la opinión pública': *El Criterio Espiritista*, 1 (1896), 15–16.
52 Henry S. Tanner, *The Human Body: A Volume of Divine Revelations Governed by Laws of God's Ordaining, Equally with Those Written on 'Tables of Stone' or in the Bible* (Long Beach, Calif., 1908).
53 Francis Gano Benedict, *A Study of Prolonged Fasting* (Washington, D.C.: Carnegie Institution of Washington, 1915).
54 Gunn, *40 Days without Food*, 99–100.

55 Although there were rumours of Tanner having committed suicide in London in May 1893 by ingesting a large dose of morphine, more reliable biographical data place his death in 1918, in San Diego, where he spent the last ten years of his life after a period in Los Angeles. See Linda B. Hazzard, *Scientific Fasting: The Ancient and Modern Key to Health* (New York: Grant Publications, 1927).

56 Barbara Gronau and Alice Lagaay (eds.), *Ökonomien der Zurückhaltung. Kulturelles Handeln zwischen Askese und Religion* (Bielefeld: Transcript Verlag, 2010).

57 Elizabeth Neswald, 'Food Fights: Human Experiments in Late Nineteenth-Century Nutrition Physiology', *Clio Medica*, 95 (2016), 170–193.

58 Charles Richet, *Physiologie. Travaux du laboratoire de M. Charles Richet. Chimie physiologique. Toxicologie* (Paris: F. Alcan, 1893–1895).

59 Jutta Person, 'Abnormität und Irrsinn – Das Spektakel des Hungerkünstlers Succi', in Torsten Hahn, Jutta Person, and Nicolas Pethes (eds.), *Grenzgänge zwischen Wahn und Wissen. Zur Koevolution von Experiment und Paranoia 1850–1910* (Frankfurt am Main: Campus, 2002), 240–254.

60 Luciani, *Fisiologia del digiuno*, 13, 16.

61 'Nous ne prétendons pas que Succi soit un fou, non plus que notre excellent Merlatti, mais, à coup sûr, ce sont des *excentriques*, des *nerveux*, des *cérébraux*, qui, soutenus par une foi exagérée et *hystérique* dans leurs forces, suppriment chez eux la sensation de faim et résistent l'inanition par un de ces phénomènes *d'autosuggestion*': Ernest Monin and Philippe Maréchal, *Stefano Merlatti. Histoire d'un jeûne célèbre, précédée d'une étude anecdotique, physiologique et médicale sur le jeûne et les jeûneurs* (Paris: C. Marpon et E. Flammarion, 1887), 2–3 (my emphasis).

62 '[C]i troviamo veramente dinanzi un pazzo, o piuttosto uno di quei tipi squilibrati che entrano a comporre il largo strato intermedio tra la saviezza e la pazzia?': Luciani, *Fisiologia del digiuno*, 15. Succi stayed in the Lungara asylum from 21 January 1883 to 4 April 1883 and from 23 November 1885 to 30 May 1886.

63 '[S]empre in uno stato di relativo miglioramento, essendo la sua malattia difficilmente sanabile': ibid., 16.

64 Cesare Lombroso, 'I digiunatore e la scienza moderna', *Il Corriere della Sera*, 2 Jan. 1887.

65 '[P]or partidario que sea del método científico o positivo, no puede dejarse de reconocer que, de un lado, es imposible explicar por las solas leyes de la física y de la química los fenómenos somáticos específicamente vitales, y de otro que los fenómenos psíquicos (de sensibilidad y conciencia), que constituyen el punto culminante de la vida, se sustraen de hecho a toda explicación mecánica … la fisiología … no puede – prescindir de considerar los fenómenos psíquicos': Luigi Luciani, *Tratado didáctico de fisiología humana* (Barcelona: Antonio Virgili, 1901), v–vi.

66 'Ce matin, le sujet a eu une attaque de nerfs qui a duré environ quinze minutes, et dont les caractères étaient ceux de l'attaque *hystérique*. Cette crise a été suivi d'un accès de dyspnée assez intense: les yeux étaient injectés et la face tuméfiée. Une application de plaques dynamométriques a eu rapidement raison de ces accidents et a démontré que Merlatti, comme tous les

néuropathes, était justiciable au plus haut point de la *métallothérapie*. Merlatti a prétendu que dans ses jeûnes précédents il avait déjà subi des semblables attaques': Monin and Maréchal, *Stefano Merlatti*, p. 183.

67 'Sans vouloir prétendre que Succi est réellement fou, c'est, à coup sûr, un excentrique, soutenu par une foi exagéré dans des forces, et peut-être aussi, dissent des interviewers, par cette auri sacra fames, qui est un stimulant de première force': Ernest Monin, 'Succi', *La revue des journaux et des livres*, 16 Oct. 1886, 776.

68 'O Succi!/Tu metti i tuoi guardiani negl'impici/Per menarteli teco a Castelpulci': Luciani, *Fisiologia del digiuno*, 123. Castelpulci was the popular name of a Florentine psychiatric hospital.

69 'La faim est une fonction tout animale dans laquelle l'esprit ne joue aucun rôle, or, comme chez les animaux, la mort arrive fatalement en assez peu de jours dans le cas d'inanition, il nous paraît impossible qu'il en soit autrement chez l'homme': quoted in Albert de Rochas, *La suspension de la vie* (Paris: Dorbon-ainé, 1913), 27–28. See also Stewart Wolf, *Brain, Mind, and Medicine: Charles Richet and the Origins of Physiological Psychology* (London: Transaction Publishers, 2012).

70 From 1892, Gley was the chair of general physiology at the museum. He also directed the *Journal de physiologie et de pathologie générale.*

71 Marcel-Eugène Gley, 'Le jeûne et les jeûneurs', *Revue scientifique*, 38 (1886), 723.

72 This was also Luciani's main result in 1888, after a one-month experimental study of Succi in his laboratory in Florence.

73 *Pall Mall Gazette*, 26 Feb. 1890, 2 (my emphasis).

74 'Batlandier à Vesoul, Savonay à Alger, Alexander Jacques à Londres, Simon à Bruxelles, jeûnèrent plus ou moins longtemps et admirent, moyennant payement, le public à les contempler; mais les recettes furent maigres, et c'est à peine si l'on parla d'eux. Il en fut de même pour un Italien, Alberto Montazzo, qui avait offert de se soumettre à une expérience de six mois': Rochas, *La suspension de la vie*, 29.

75 *La Nature*, 20 Nov. 1886. Rochas's papers are kept at the archive of the American Philosophical Society Library, Mss. Ms. Coll. 106.

76 Fonvielle, *Mort de faim. Étude sur les nouveaux jeûneurs* (Paris: Librairie illustrée, 1886), 286.

77 'Sa conviction neutralise la sensation de faim': 'Une hypothèse', *La revue des journaux et des livres*, 6 Mar. 1886, 319.

78 'Les recherches récemment faites sur l'hypnotisme ont jeté de curieuses lumières sur le problème de la suspension plus ou moins partielle des fonctions vitales': Monin and Maréchal, *Stefano Merlatti*, 37. See also Hippolyte Bernheim, 'Physiologie pathologique. Le jeûne de Succi', *Gazette hebdomadaire de médicine et de chirurgie*, 42 (1886), 681–683.

79 Hippolyte Bernheim, *De la suggestion et de ses applications à la thérapeutique*, 2nd ed. (Paris: O. Doin, 1888).

80 Dominique Barrucand, 'Freud et Bernheim', *Histoire des sciences médicales*, 20 (1986), 157–170; Nieto-Galan, 'Useful Charlatans'.

81 '[A]yant suggéré à deux femmes hystériques endormies par lui l'absence de faim et l'ordre de ne pas manger, [il] put les soumettre à un jeûne de quinze

jours pleins, pendant lesquels elles ont bu, mais n'ont ingéré aucun aliment solide. Ce jeûne, très bien supporté, aurait pu être prolongé encore pendant quinze jours, mais l'une des malades avait déjà perdu 3 k. 200 et l'autre 5 k. 200': Bernheim, *De la suggestion*, 37.

82 Like opium, morphine, or chloroform.

83 Hippolyte Bernheim, 'Le jeûne de Succi', *Journal des connaissances médicales*, 54 (1886), 341–343.

84 Monin and Maréchal pointed out that 'La conclusion du professeur Bernheim est donc que Succi est un croyant, neutralisant la sensation faim, fanatisé qu'il est par sa foi dans l'efficacité de son breuvage, il ne meurt pas de faim parce qu'il n'a pas faim': Monin and Maréchal, *Stefano Merlatti*, 78. During Succi's fast in Barcelona in 1888, Dr Badía used hypnotism as another possible explanation for the extraordinary resistance to inanition shown by professional fasters.

85 Charles Richet, 'Physiologie. Course de physiologie de la Faculté de Médecine de Paris. L'inanition chez l'homme', *Revue scientifique*, 4 (1889), 106–112; 26 (1889), 801–805.

86 Ibid., 107. Richet paid particular attention to the impressive resistance to inanition of hysterics and maniacs.

87 'Un homme sain meurt, s'il cesse de manger après un certain nombre de jours; un homme malade reste impunément plusieurs semaines sans se nourrir': Bernheim, 'Le jeûne de Succi', 342.

88 Charles Richet, 'La Faim', *Dictionnaire de Physiologie*, 6 (1904), 10; Benedict, *A Study of Prolonged Fasting*, 213.

89 *Pall Mall Gazette*, 26 Feb. 1890, 2.

90 'Ups and Downs of Succi's Spell', *New York Herald*, 21 Dec. 1890.

91 'Le célèbre jeûner Succi dont on a tant parlé, est devenu subtilement fou. On a dû l'enfermer à l'asile Sainte-Anne et lui mettre la camisole de force. Est-ce aux jeûnes prolongés que Succi doit sa folie?': *La Gazette du village*, 28 (1891), 268.

92 'Yo creo que estoy curado y reconozco que he estado loco. La causa tiene probablemente dos caracteres. Durante seis meses que pasé en Londres, no comí casi nada, y bebí té solamente. No bien llegué a París, empecé a beber un poco de todo y concluí por emborracharme en más de una ocasión … Un amigo mío hipnotizador vino conmigo – desde Londres y se divertía en hipnotizarme … Puede ser que haya en mi propensión a la locura el mismo poder que tengo para ayunar durante largo tiempo': *El Universal de México*, 31 Jul. 1892, 1.

93 Other newspaper articles were reluctant to accept Succi's own thesis, and his case was considered by many as 'A lesson … taught to all professional fasting men and women by the fate of Succi': *Huntly Express*, 4 Jun. 1892, 7. Nevertheless, in 1897, the *Press* reported that: 'Succi, the Italian fasting man, went suddenly mad last night … he had barely entered the House before he seized a stick, and, shouting out at the top of his Voice, started breaking everything in the room': *Press*, 22 Feb. 1897, 4.

94 Heather Wolffram, 'Hallucination or Materialization? The Animism versus Spiritism Debate in Late Nineteenth-Century Germany', *Journal of the History of the Human Sciences*, 25 (2) (2012), 45–66.

95 Henri Lassignardie, *Essai sur l'état mental dans l'abstinence* (Bordeaux: Imprimerie du Midi, 1897).
96 *Belfast Telegraph*, 15 Jun. 1905, 4 (my emphasis).
97 Benedict, *The Influence of Inanition on Metabolism*, 3.
98 Émile Desportes, *Du refus de manger chez les aliénés* (Paris: Parent, 1864).
99 Benedict, *The Influence of Inanition on Metabolism*, 3.
100 Morgulis, *Fasting and Undernutrition*, 8.
101 Ibid., 9.
102 'Miss Emmey. Esta célebre ayunadora, émula de los Merlatti, Succi, Tanner, Pappus, se presentará al público de Madrid el domingo 19 en el Frontón Central para hacer sus interesantes experimentos. A las diez de la noche del mencionado dia, Miss Emmey entrará en una urna de cristal, que será sellada después y en la cual permanecerá por espacio de siete días sin tomar alimento alguno. Durante dicho tiempo, podrá ser visitada por el público a cualquier hora del día o de la noche': *El País*, 16 Oct. 1902, 3.
103 Sigal Gooldin, 'Fasting Women, Living Skeletons and Hunger Artists: Spectacles of Body and Miracles at the Turn of a Century', *Body and Society*, 9 (2) (2003), 37. Ann Moore, for example, was supposedly fed by her daughter in several ways; Margaret Weiss, another fasting girl, only ten years of age, had such powers of deception that 'she grew, walked, and talked like other children of her age, still maintaining that she used neither food nor drink. In several other cases reported all attempts to discover imposture failed. As we approach more modern times the detection is more frequent': George M. Gould and Walter L. Pyle, *Anomalies and Curiosities of Medicine* (Philadelphia: W. B. Saunders, 1897), 418.
104 'Like Kafka's hunger artists defeated by viewers' indifference to his sacrifice, the eventual absence of sympathetic onlookers would reduce fasting once again to pathology – and a feminized one at that – as anorexia nervosa moved ever more insistently into public spotlight, erasing public memory of the bygone apostles of abstinence': R. Marie Griffith, 'Apostles of Abstinence: Fasting and Masculinity during the Progressive Era', *American Quarterly*, 52 (4) (2000), 630.
105 Nina Diezemann, *Die Kunst des Hungerns. Essstörungen in Literatur und Medizin um 1900* (Berlin: Kadmos, 2006).
106 Walter Vandereycken and Ron van Deth, *From Fasting Saints to Anorexic Girls: The History of Self-Starvation* (London: Athlone Press, 1990), 87.
107 Griffith, 'Apostles of Abstinence', 603.
108 Barbara Gronau, 'Asceticism Poses a Threat: The Enactment of Voluntary Hunger', in Alice Lagaay and Michael Lorber (eds.), *Destruction in the Performative* (Amsterdam: Rodopi, 2012), 99–109.
109 Vandereycken and van Deth, *From Fasting Saints to Anorexic Girls*.
110 Robert Fowler, *A Complete History of the Welsh Fasting Girl Sarah Jacob* (London: Henry Renshaw, 1871).
111 'There must be something unnatural in the mental training of those individuals who, in this nineteenth century, announce themselves as patrons of the miraculous and abnormal! ... It is in these days rare to find medical men amongst the votaries of marvels. Modern Medicine, guarded by the

simplicity of the Dissecting-room and the exactitude of the Laboratory, is, from the observation of things *without*, led up by the laborious process of induction, to a more correct appreciation of Nature's unerring laws': ibid., 2. See also Siân Busby, *A Wonderful Little Girl: The True Story of Sarah Jacob, the Welsh Fasting Girl* (London: Short Books, 2003).

112 'We have advanced the theory that – the girl had brought herself to believe that even her parents were ignorant of her deceptions and that from motives we have detailed, the parents themselves had likewise concealed their acquired knowledge from their invalid daughter': Fowler, *A Complete History of the Welsh Fasting Girl Sarah Jacob*, 258–259.

113 Sofie Lachapelle, 'Between Miracle and Sickness: Louise Lateau and the Experience of Stigmata and Ecstasy', *Configurations*, 12 (2004), 77–105.

114 Fonvielle, *Mort de faim*, 45–60.

115 Augustus Rohling, *Louise Lateau: Her Stigmas and Ecstasy* (New York: Hickey, 1876).

116 Rudolf Virchow, *Uber Wunder* (Breslau: Morgenstern, 1874).

117 Fonvielle, *Mort de faim*.

118 Gunn, *40 Days without Food*.

119 William Alexander Hammond, *Fasting Girls: Their Physiology and Pathology* (New York: G. P. Putnam's Sons, 1879). See also Joe Nickell, 'Mystery of Mollie Fancher: The Fasting Girl and Others Who Lived without Eating', *Skeptical Inquirer*, 41 (6) (2017), 18–21.

120 Gunn, *40 Days without Food*, 31.

121 Ibid., 33.

122 Ibid.

123 Ibid.

124 Some years later, at the end of the century, a paper on anomalies and curiosities of medicine described anorexia nervosa as a digestive disturbance of hysteria, with remarkable cases of survival for long periods without food, especially in women, adding again some remarks about the difficulties of achieving a complete absence of nourishment in the body. 'Abstinence, or rather anorexia, is naturally associated with numerous diseases, particularly of the febrile type, but in all these cases the patient is maintained by the use of nutrient enemata or by other means, and the abstinence is never complete': Gould and Pyle, *Anomalies and Curiosities of Medicine*, 413.

125 Hazzard, *Scientific Fasting*.

126 Griffith, 'Apostles of Abstinence', 603.

127 'La ayunadora de París: Dicen A *El Imparcial* desde París que la ayunadora Miss Alma Nelson continúa siendo la curiosidad del día. Envuelta en una bata azul celeste, relata a sus numerosos visitantes la historia de su vida y de su descubrimiento. Nació en Nueva-York, de padres pobres, que muy niña la dejaron huérfana, entregada a su nodriza, una india brava que conocía, por su vida semisalvaje, la virtud secreta de muchas plantas. Alma cayó un día gravemente enferma. La India la cuidó y la curó … Miss Nelson vino a París en su juventud, cantando en varios cafés-conciertos. Poseía ya el secreto de la India, y habiéndolo experimentado repetidas veces en sí misma. La ayunadora afirma que ha llegado a pasar hasta veintisiete

días sin tomar más alimento que dos vasos de la preciosa infusión, unos 60 centilitros cada veinticuatro horas. El elixir es amarillento, espeso como un jarabe y de un sabor aromático *sui generis*. Miss Nelson muéstrase orgullosa de su ensayo, que cree mucho más útil que las apuestas de Merlatti y de Succi. Varios médicos la visitan cotidianamente. Un enfermero la vigila día y noche. Un boletín colocado a la puerta de su habitación marca el estado del pulso, de la temperatura y de la respiración. Hasta ahora no ofrecen alteración alguna extraordinaria': *La Época*, 31 Jan. 1892.

128 Benedict, *The Influence of Inanition on Metabolism*, 6; Gronau, 'Asceticism Poses a Threat'.

129 '[D]as schwache Geschlecht einen starken Magen [habe]', *Illustrirtes Wiener Extrablatt*, 23 Jul. 1905, 8; quoted in Peter Payer, 'Hungerkünstler. Anthropologisches Experiment und modische Sensation', in Brigitte Felderer and Ernst Strouhal (eds.), *Rare Künste. Zur Kultur- und Mediengeschichte der Zauberkunst* (Vienna: Springer, 2006), 262.

130 Dr Theodor Brugsch and Dr Rahel Hirsch, 'Gesammt-N und Aminosaurennausscheidung im Hunger', *Zeitschrift für experimentelle Pathologie und Therapie*, 3 (1906), 638–645; M. Bonniger and L. Mohr, 'Die Saurebildung im Hunger', ibid., 675–687; R. Baumstark and L. Mohr, 'Ueber die Darmfaulniss im Hunger', ibid., 687–691; Benedict, *The Influence of Inanition on Metabolism*, 6.

131 Bodleian Libraries, University of Oxford, John Johnson Collection: London Play Places 1 (63).

132 Fred Dangerfield, *The Playgoer*, vol. I (London: Dawban and Ward, 1901–1902).

133 Van Hoogenhuyze and Verploegh's observations on the fasting girl Flora Tosca appeared in *Zeitschrift für physiologische Chemie*, 46 (1905), 440; see also Brugsch and Hirsch, *Zeitschrift für experimentale Pathologie und Therapie*, 3 (1906), 638; Bonniger and Mohr, ibid., 675; Baumstark and Mohr, ibid., 687.

134 Francis G. Benedict and A. R. Diefendorf, 'The Analysis of Urine in a Starving Woman', *American Journal of Physiology*, 18 (1907), 362.

135 Van Hoogenhuyze and Verploegh, 'Beobachtungen über die Kreatininausscheidung beim Menschen', *Zeitschrift für physiologische Chemie*, 46 (1905), 415; Bönniger and Mohr, 'Untersuchungen uber einige Fragen des Hungerstoffwechsels. I. Säurebildung im Hunger', *Zeitschrift für experimentale Pathologie und Therapie*, 3 (1906), 675; Benedict and Diefendorf, 'The Analysis of Urine in a Starving Woman'.

136 'A Berlin, Mlle de Serval, qui a fait de sérieuses études de médecine et qui considère la plupart de nos maladies comme dues à une nourriture trop abondante et trop irrégulière, s'est guérie de plusieurs infirmités par de courtes cures de jeûne rigoureux de deux à six jours en se contentant de boire de l'eau pure; elle estime que tout le monde (jeunes et vieux, malades et gens bien portants) devrait jeûner deux jours par semaine. Afin de permettre aux médecins d'étudier les effets physiologiques de la faim, elle s'est enfermée volontairement plusieurs fois dans une cage vitrée de 3 mètres de longueur sur 2 m. 50 de largeur et 2 mètres de hauteur, hermétiquement fermée des quatre côtés … Malgré les conditions

hygiéniques peu favorables d'une cage assez étroite pour empêcher tout exercice, toutes les fonctions physiologiques se maintenaient parfaitement normales ... Malgré la pâleur de sa peau, les médecins ont constaté, dans le cas de Mlle de Serval, une teneur du sang remarquablement constante en hémoglobine ... Le dernier jeûne et le plus long de cette dame a été de quarante jours, pendant lesquels elle s'est abstenue de toute nourriture et n'a absorbé qu'une faible quantité d'eau pure (au lieu de l'eau minérale autrefois utilisée). Les lettres écrites pendant sa captivité volontaire témoignent de la lucidité parfaite et de l'activité d'un esprit hautement cultivé': Rochas, *La suspension de la vie*, 34–35.

137 Avelino Martín, *El ayuno de Succi. Contribución al estudio de la inanición* (Barcelona: Tipografía de la Casa Provincial de Caridad, 1889), 5.

138 Ibid., 16. Similarly, Dr Antonio Riera endorsed Martín's physico-chemical reductionism, praised experiments with hunger artists, and referred in general terms to a longstanding materialist tradition in physiology. See Javier de Benavent, *Ayuno Succi* (Barcelona: Tipolitografía de Busquets y Vidal, 1890), 22.

139 Dr Raimundo Comet and Dr Francisco Carbó defended a strict physico-chemical explanation for the problems, the latter admitting, however, that nutrition was at that time one of the most obscure branches of physiology. See Estanislao Andreu Serra, 'Memoria sobre los trabajos, estado y progreso de la Academia Médico Farmacéutica de Barcelona', *La Enciclopedia*, 1 (1888), 56–58. 'A las ocho y media de esta noche celebrará sesión pública ordinaria la Academia Médico-farmacéutica [de Barcelona], en la que dará cuenta de las observaciones hechas sobre el ayuno de Succi, el señor Don Andrés Avelino Martín': *La Vanguardia*, 4 Dec. 1888, 2. The journal of the Academia was *Enciclopedia Médico-Farmacéutica.*

5 Elixirs

1 John Duffy, *From Humors to Medical Science: A History of American Medicine* (Urbana: University of Illinois Press, 1993). See also Henry S. Tanner, *The Human Body: A Volume of Divine Revelations, Governed by Law of God's Ordaining, Equally with Those Written on 'Tables of Stone' or in the Bible* (Long Beach, Calif., 1908), 21.

2 Tanner considered that 'too much importance was given to the action of food both in health and disease, and not enough to good air and water': R. A. Gunn, *40 Days without Food: A Biography of Henry S. Tanner MD, Including a Complete and Accurate History of His Wonderful Fast, 42 Days in Minneapolis, and 40 Days in New York City, with Valuable Deductions* (New York: Albert Metz, 1880), 9.

3 'Unable to disentangle themselves from their botanic biases and ill-equipped to accept the implications of germ theory, the financial costs of salaried faculty and staff, or the research implications of laboratory science, the eclectics emerged into the twentieth century stubborn and fundamentally perplexed by the rigor demanded in the new medical

curriculum and in the expectations of research. By the time Abraham Flexner wrote his landmark *Medical Education in the United States and Canada* (1910), the school's remnants were all but forgotten in the rush of modern academic medicine': John S. Haller, *Medical Protestants: The Eclectics in American Medicine, 1825–1939* (Carbondale: Southern Illinois University Press, 1994), xix.

4 Duffy, From *Humors to Medical Science*.

5 Laurence Brockliss and Colin Jones, *The Medical World of Early Modern France* (Oxford: Clarendon, 1997); Michael Bycroft, 'Iatrochemistry and the Evaluation of Mineral Waters in France, 1600–1750', *Bulletin of the History of Medicine*, 91 (2) (2017), 303–330.

6 Penelope J. Corfield, 'From Poison Peddlers to Civic Worthies: The Reputation of the Apothecaries in Georgian England', *Social History of Medicine*, 22 (1) (2009), 1–21. See also Anne Digby, *Making a Medical Living: Doctors and Patients in the English Market for Medicine, 1720–1911* (Cambridge: Cambridge University Press, 1994), 7.

7 Ann Anderson, *Snake Oil, Hustlers and Hambones: The American Medicine Show* (Jefferson, N.C.: McFarland, 2000); Christopher R. De Corse, 'Elixirs, Nerve Tonics and Panaceas: The Medicine Trade in Nineteenth-Century New Hampshire', *Historical New Hampshire*, 39 (1–2) (1984), 1–23.

8 For a compilation of contemporary drugs, see Giuseppe Orosi, *Manuale dei medicamenti galenici e chimici, con la descrizione dei loro caratteri, la loro preparazione, la virtù terapeutica, le formule di uso medico, le incompatibilità relative, le adulterazioni commerciali, gli antidoti ecc.* (Florence: Eugenio e F. Cammelli, 1867). The publisher was located in Piazza della Signoria, not far from the printing house where Succi printed his *Il Corriere Spiritico*.

9 Jeremy Agnew, *Entertainment in the Old West: Theater, Music, Circuses, Medicine Shows, Prize Fighting and Other Popular Amusements* (Jefferson, N.C.: McFarland & Company, Inc., 2011); De Corse, 'Elixirs, Nerve Tonics and Panaceas'.

10 In the 1880s, pharmacist Charles Alderton created Dr Pepper, a carbonated soft drink with digestive properties.

11 Allison C. Meier, '15 Curious Quack Remedies from the Age of Patent Medicine', www.mentalfloss.com/article/85554/15-curious-quack-remedies-age-patent-medicine (accessed 10 June 2020). The fifteen are: opium, blood, cocaine, prairie flowers and Indian oil, petroleum, cannabis, tomatoes, arsenic, hair tonics, radioactive substances, mercury, obesity bath powder, swamp root, Dr Pepper, and pink pills for pale people.

12 Roy Porter and Mikulas Teich (eds.), *Drugs and Narcotics in History* (Cambridge: Cambridge University Press, 1995).

13 Robert Jütte, 'The Paradox of Professionalism: Homeopathy and Hydropathy as Unorthodoxy in Germany in the Nineteenth and Early Twentieth Century', in Robert Jütte, Guenter B. Risse, and John Woodward (eds.), *Culture, Knowledge and Healing: Historical Perspectives of Homeopathic Medicine in Europe and North America* (Sheffield: EAHMH, 1998), 65–68.

14 Such as Salvador Badía, Javier de Benavent, Joaquim Homs, and Ramón Roig.

15 John Haller, *The History of American Homeopathy: From Rational Medicine to Holistic Health Care* (New Brunswick, N.J.: Rutgers University Press, 2009).
16 Anti-diarrhoeal, taeniafuges, and a tonic called 'osteógeno': Joel Piqué, 'Procesos de construcción social y científica de la homeopatía en Catalunya (1890–1924)', PhD Thesis, Universitat Autònoma de Barcelona, Bellaterra, 2018, 161.
17 *El Diluvio*, 24 Nov. 1888.
18 Javier de Benavent, *Ayuno Succi* (Barcelona: Tipo-Litografía de Busquets y Vidal, 1890), 22.
19 '[D]urante la salud, una fuerza spiritual rige al organismo, manteniendo en él la armonía, ya que sin esta fuerza vital o espiritual, el organismo no puede vivir, ya que en la enfermedad, la fuerza vital está sólo desarmonizada primitivamente de un modo morboso, expresando su sufrimiento por anomalías en el modo de obrar y de sentir del organismo, pues para cursar una enfermedad cualquiera, ya sea física, ya moral, es inútil saber cómo la fuerza vital produce los síntomas … los médicos y hasta los que no lo son, de ideas materialistas o positivistas, no llegan a comprender esto, como tampoco comprenden el modo de obrar de *nuestros medicamentos casi siempre infinitesimales,* acostumbrados como están a no ver más que fuerzas ponderables a materiales en todo': Javier de Benavent, *Propaganda homeopática* (Barcelona: Imprenta de Collazos y Tasis, 1897), 12 (my emphasis).
20 Jütte, Risse, and Woodward (eds.), *Culture, Knowledge and Healing*, 1–4.
21 Wilhelm Ameke, *History of Homeopathy: Its Origin, Its Conflicts. With an Appendix of the Present State of University Medicine* (London: E. Gould and Son, 1885).
22 'Highly poisonous substances could thus be converted into beneficent and innocuous remedies, and substances which were easily decomposed, and therefore tending to become inefficacious, could be converted into a form in which they were not liable to decomposition, and thereby become powerful remedial agents in the hand of a skilful physician': ibid., 131.
23 Salvador Badía, 'El ayunador Succi en Barcelona. Decálogo homeopático', *El Consultor homeopático*, 13 (1888), 354–358; 14 (1888), 378–380. I am indebted to Ms Neus Marin and to Dr Encarna Villar of the Acadèmia Mèdico Homeopàtica de Barcelona for providing me with access to this very valuable article. Members of the editorial board of *El Consultor Homeopático* were Salvio Almató, Salvador Badía, Javier de Benavent, Manuel Cahís, and José Nogué.
24 '[P]ara nosotros es un hombre extraordinario que está resolviendo uno de las más interesantes problemas de la Terapéutica; que es poder sostener la individuo por cierto tiempo para poder curar muchas enfermedades del estómago, del hígado y de los centros de nutrición que necesitan de abstinencia para poder lograr una curación … lo que … tiene el experimento de Succi de notable es que se trata de un fenómeno fisiológico llevado a cabo sin que entre en el terreno de la patología': ibid., 354, 355.
25 Badía considered that Succi's experiment was useful for the: 'clarification of the mechanism of digestion, [and] … to open up a new field in the therapeutics of diseases which require a strong depletion for which we need a certain diet … and in which absolute rest of the stomach is needed': ibid., 356.

26 See http://dbe.rah.es/biografias/37945/anastasio-garcia-lopez (accessed 23 Jun. 2020); Julián Martín Oliver, 'El Instituto Homeopático y Hospital de San José de Madrid y su entorno profesional en el último tercio del siglo XIX', PhD, Universidad Complutense, Madrid, 2014. See also Cristina Albarracín Serra, 'Homeopatía y espiritismo. La obra del Dr Anastasio García López', master's thesis, Universidad Complutense, Madrid, 1988. García López believed that they could perfectly match other medical practices such as hydropathy (with different mineral waters), vegetarianism, and obviously homeopathy: *Del estado de la doctrina homeopática y de su porvenir en la ciencia* (Madrid: Imprenta de M. Rivadeneyra, 1868); *Tratado de hidrología médica, con La guía del bañista y el Mapa balneario de España* (Madrid: Carlos Bailly-Baillière, 1869); *Cartas críticas sobre la medicina y los médicos* (Salamanca: Imprenta de Sebastián Cerezo, 1871); *Lecciones de medicina homeopática* (Madrid: Imprenta de M. Rivadeneyra, 1872); *Refutación del materialismo* (Madrid: Imprenta de Alcántara, 1874); *Hidrología Médica* (Salamanca: Imprenta de Sebastián Cerezo, 1875).

27 See http://dbe.rah.es/biografias/37945/anastasio-garcia-lopez (accessed 23 Jun. 2020).

28 Jütte, 'The Paradox of Professionalism', 65.

29 In Germany in 1860, homeopathy was flourishing, with 259 registered practitioners, 2 national societies, 6 state societies, 3 local societies, 13 homeopathic hospitals, 14 homeopathic apothecaries, 4 homeopathic medical journals, and 4 homeopathic popular journals: ibid., 72.

30 *Globe*, 17 Dec. 1891, 2 (my emphasis).

31 'Certains états constitutionnels favorisent le jeûne et la résistance des êtres à une carême prolongée. On a fait grand bruit, il y a un certain nombre d'années, autour de Tanner, de Merlatti et de Succi, qui se sont rendus célèbres, l'un par un jeûne de quarante jours, les autres par un jeûne de cinquante jours': 'Des maladies de l'estomac': *La Clinique. Organe de l'Homeopathie complexe*, 257/258 (1896), 26.

32 '[D]irigirme, en un lenguaje comprensible, al vulgo de las gentes, que es donde únicamente echa raíces el sistema homeopático, porque la savia suministrada por la ignorancia es la sola que conviene a su nutrición. En breves párrafos procuraré pues, principalmente rebatir lo que más de una vez he oído hacer a los partidarios de la homeopatía, en defensa de lo que ellos llaman Sistema cómodo de curar las enfermedades. En cuanto me sea posible, he de colocarme para ello en los límites de la verdad científica, de tal manera que, a la vez que combatir el sistema homeopático, he de señalar también los peligros de la polifarmacia': Luis C. Maglioni, 'Homeopatía. Tesis para optar al Doctorado', Facultad de Ciencias Médicas, Buenos Aires, 1877, 10–11.

33 Dr Luis C. Maglioni, *Mis 37 días de ayuno (auto-experiencia). A propósito de los casos de Mr. Mac Sweeney y sus 11 compañeros, ayunadores políticos en las prisiones de Brixton y de Cork. 'Huelga de Hambre'* (Buenos Aires: Imprenta San Martín, 1920), 47.

34 'Invece di mangiare egli ha bevuto liquor, che non vi sono all'osteria. Digiunatore di ferro! Io ti saluto': Giovanni Chiverny, *Del Signor Succi e del suo digiuno* (Milan: Tipografia di Giacomo Pirola, 1886), 121.

35 In 1895, for example, Dr John Rees, professor of toxicology in Philadelphia, again brought to the fore the tension between food and drugs in the following terms: 'Sr Succi has a liquid, which he calls "*Elixir Medicinale Succi*"; this, he claims, is a "secret" which was given to him by certain tribes in Africa ... He takes 15 to 40 minims of this occasionally, frequently going 4 to 5 days without taking any, and denies that there is any nutriment in it; but says he only takes it to alleviate intestinal pains. A *chemical analysis of this liquid shows it to contain morphine hydrochloride, chloroform, ether, cannabis indica and alcohol* [...] Dr Eisner [...] states that the mysterious "Elixir", of which Succi has occasionally partaken, consists of *"kola" (an African bean, possessing properties somewhat similar to those of coca), camphor, morphine and valerian*': John Rees, *Textbook of Medical Jurisprudence and Toxicology*, 4th ed. (Philadelphia: Blakiston's Son and Co., 1895), 189–191 (my emphasis).

36 Barbara Hodgson, *In the Arms of Morpheus: The Tragic History of Laudanum, Morphine, and Patent Medicines* (Buffalo, N.Y.: Firefly Books, 2001).

37 For other travels to Africa, see Georges Revoil, *La Vallée du Darror, voyage aux pays çomalis (Afrique orientale)* (Paris: Challamel aîné, 1882).

38 '[P]ains d'opium brut dit kassoumba ... Grâce à cet opium, les estomacs ne crient plus famine dans les moments critiques': Ernest Monin and Philippe Maréchal, *Stefano Merlatti. Histoire d'un jeûne célèbre, précédée d'une étude anecdotique, physiologique et médicale sur le jeûne et les jeûneurs* (Paris: C. Marpon et E. Flammarion, 1887), 75–76.

39 *Les annales politiques et littéraires*, 12 Sep. 1886, 165. *Osmazone* is a Greek word meaning meat flavour or broth flavour, like fibrin, obtained from animal extracts, probably consisting of one or several substances.

40 'Le secret Succi fait travailler fort inutilement des cervelles. Nous avons eu la cocaïne, l'arsenic, l'extrait concentré de concombre, aujourd'hui c'est la *Carica Papala*, fruit d'une plante de la famille des cucurbitacées, la seule des végétaux qui contienne de l'osmazone, la substance nutritive de la viande': *La petite presse*, 6 Sep. 1886 (my emphasis).

41 'En 1877, il eut des fièvres d'Afrique et s'aperçut à ce moment que certains sucs végétaux qu'il prenait pour combattre ces fièvres lui permettaient de s'abstenir de toute nourriture': Luigi Buffalini, 'Les résultats d'un jeûne prolongé', *La semaine médicale*, 19 Sep. 1886.

42 '[E]spèce de marron astringent, très apprécié des peuples de l'Afrique centrale pour ces propriétés reconstituantes, et permettant aux voyageurs de supporter sans fatigue la privation de nourriture et des longues marches sous un soleil énervant': ibid., quoted by Monin and Maréchal, *Stefano Merlatti*, 67.

43 L'abbé Vallée, curé des Monts, 'Essai d'explication théorique et scientifique du jeûne extraordinaire des italiens Succi et Merlatti', *Annales de la Société d'agriculture, sciences, arts et belles lettres d'Indre-et-Loire*, 67 (1887), 32.

44 'Der Succi-Liqueur/Der Succi, der grosse Gelehrte/Erfanden famosen liqueur/Und wer davon trinkt nur ein wenig/Den hungert's ganz sicher nicht mehr/Man sehnt sich nach Schnitzel und Roastbeef/Nach Gollasch und Beuschel nichts mehr/Und auch nicht nach Nudeln und Strudel/Bei diesen brillanten liqueur': *Wiener Caricaturen*, 25 Jul. 1886, 2.

45 'Signor Succi is a traveller by profession ... [who] has succeeded in attracting the attention of the medical faculty ... It is stated that the elixir upon which

Succi claims to support life consists simply of an *arsenious solution*. It is well known that arsenic possesses the property of producing in the stomach a warmth, which temporarily allays the pangs resulting from the deprivation of nourishment. The chemist who has suggested this view of the question is said to purpose experimenting upon himself with an arsenious solution': *Chemist and Druggist*, 18 Sep. 1886, 391.

46 'De che mi servo per tenere unito il pensiero e mantenere le forze senza nutrizione digestiva?; De che mi servo per attuire la sensazione dolorosa della fame nei primi sete giorni?; Che cosa adopero per unire il veleno alla materia e renderlo omogeneo allá estesa?': 'Notizie Diverse', *Il Corriere Spiritico*, 2 (1888), 68.

47 'Liquore Succi. Firenze. Piazza della Signoria, 7 (Angolo Via dei Magazzini). Il liquore Succi è medicinale ed è composto di sostanze vegetali d'Africa e fu usato dall'Inventore nei suoi viaggi in Africa Equatoriale. È medicamento portentoso di azione tonica e sotto tutti i rapporti profilatico e perciò di subitaneo effetto per le malalttie seguenti: Colera, Dissenteria, Diarrea, Coliche, ecc. ecc. Emicranie persistenti, Vertigini, Isterismo, Corea, Neuralgia, Tetano, Eclampsia (dolori delle partorienti), Reumatismo, Dispepsia, Acidità, Dolori, Atonia intestinale, Epilessia, Asma, Delirio dei bevitori, Vizio cardiaco, Melanconia, ecc. È un eccellente preservativo delle Febbri di Malaria er è efficace calmante ed eccitante. Si adopera pure per le persone che soffrono di insonnia e di mal di mare. E un eccellente tonico usandone tre o quattro gocce con vino o vermouth. Questo Liquore è uno dei piú eccellenti antispasmodico, tant è vero che nei primi gironi del diggiuno il Succi se ne serve per attutire la sensazione dolorosa della fame, come chiaramente fu dimostrato nei diverse experimenti da esso fatti, fra i quali nomineremo quelli del Cairo d'Egitto, Forli, Milano, Parigi e per ultimo di Firenze fatto nel Marzo 1888 sotto la soverglianza e gli studii dell'Accademia Medico Fisica': *Il Corriere Spiritico*, 1 (6) (1888), 160.

48 'I. Forma ottagonale di una boccia su cul trovansi rilevate sul vetro le parole Liquore Succi
II. Timbro e ceralacca rossa con impresse le parole: Succi-Firenze.
III. Timbro a umido, rosso, colla iscrizione: Liquore Medicinali Succi Firenze, Piazza della Signoria, 7.
IV. Timbro a umido con fac-simile della firma: G. Succi
V. Etichetta bianca colla scritura: Liquore del Digiunatore, Esploratore Giovanni Succi ecc.
VI. Foglio da involto nel cui fondo havvi il ritratto del Succi fra due leoni e la stritta: Liquore ecc.
VII. Striscia di carta verde collo stesso emblema e colla scrita G. Succi ecc.

Detto marchio o segno distintivo di fabbrica sarà del richiedente adoperato a contraddistinguere lo speciale liquore da lui fabgricato e intitolato sal suo proprio nome.'

See *Gazzetta Ufficiale del Regno d'Italia*. Roma. Supplemento al numero 130, 2 Jun. 1888, 298–299.

49 *La época*, 29 Nov. 1886.

50 *The People*, 7 Nov. 1886 (my emphasis).

51 Noel Paton and Ralph Stockman, *Proceedings of the Royal Society of Edinburgh*, 16 (1888–1889), 121. 'Jacques ... took a powder made of herbs to which he naturally attributed his power of prolonging life without food': George M. Gould and Walter L. Pyle, *Anomalies and Curiosities of Medicine* (Philadelphia: W. B. Saunders, 1897), 421.

52 *La época*, 29 Nov. 1886.

53 Monin and Maréchal, *Stefano Merlatti*, 54–56.

54 'On a fait toutes les hypothèses sur sa composition': Ernest Monin, 'Succi', *La revue des journaux et des livres*, 16 Oct. 1886, 775.

55 'Le maté est la vraie boisson alimentaire des climats débilitants. Ses propriétés toniques et excitantes permettent de supporter aisément un jeûne prolongé. Il est probable que c'est son infusion qui imbibait la fameuse serviette humide du docteur Tanner. Le maté trompe la faim de l'indien et du sud-américain': Ernest Monin, *L'hygiène de l'estomac. Guide pratique de l'alimentation* (Paris: O. Doin, n.d.), 329.

56 'La fameuse liqueur n'est qu'une mystification, pour amuser les badauds, ou mieux pour servir de base à des réclames futures': Monin, 'Succi', 775.

57 'LUCIANI: Io credo che questa non potrà mancarvi in seguito, se avrete giudizio e se terrete conto dei consigli della scienze. Che bisogno c'è per esempio, di lasciar credere che la vostra resistenza au digiuni sia merito del famoso liquore, mentre invece è chiaro che il merito è tutto vostro?

SUCCI: Senta, io non ho mai preteso che il fenomeno dipenda esclusivamente dal mio liquore. Se i giornali l'hanno affermato, io non me ne impaccio. A Lei dirò francamente che il liquore non mi serve ad altro che a risparmiarmi i dolori di stomaco nei primi due giorni di digiuno. L'agente principale nelle mie esperienze è una forza più potente ...

[...]

LUCIANI: Ma insomma che cosa volete dimostrarci voi col digiuno che vi proponete di ripetere?

SUCCI: Io propongo a voi scienziati due domande ...

LUCIANI: Quesiti ...

SUCCI: Dico due domande, e sono le seguenti: *Come si spiega la mia capacità di unire veleno alla materia, e far sì che il veleno divenga omogeneo alla stessa? Come si spiega che io riesco a tener unito il mio pensiero al mio corpo, e a mantenere le mie forze senza la nutrizione digestiva?* (Parole testualli).

LUCIANI: Sono problemi difficili ... non si può negare. Ma voglio sperare che voi non vi rifiuterete a metterci sulla via per risolveri. Dite, avreste difficoltà di fornirmi una certa dose del vostro veleno che immagino sia rappresentato dal liquore ...

SUCCI: Nessuna dificoltà. Ma La prevengo che se Lei vuole sperimentarlo sui cani, no ci capirà niente, perchè lo sopportano benissimo anche a forte dose.

LUCIANI: Ah sí? ... Allora mi risparmierò la fatica di sperimentarlo.

SUCCI: Ma non creda che sia tutta qui la mia tolleranza. Al Cairo ho fatto delle prova di forza da far sbalordire. Fui buono di prendere tutto d'un colpo tale quantità di morfina e di laundano, che i miei amici rimasero sgomentati, ed erano persuasi che ne sarei morto. Anche a Parigi, come risulta da un document che le mostrerò ...

LUCIANI: Ascoltate: la prova più persuasiva che potreste fornirmi di cotesta vostra tolleranza ai veleni, sarebbe di ripetere con noi un esperimento.

SUCCI: Ora non mi conviene, caro Professore. Voglio limitarmi in questa occasione al semplice digiuno di 30 giorni. La questione dei veleni ... mi riserverei a trattarla in altra occasione.

LUCIANI: Vi parrebbe di mettere ora troppa carne al fuoco.

SUCCI: Ecco, precisamente.'

See Luigi Luciani, *Fisiologia del digiuno: studi sull'uomo* (Florence: Tipografia dei Succesori Le Monnier, 1889), 25–26.

58 As Luciani pointed out: 'the sensation of hunger may be tolerable in the first two days of abstinence, and may decrease and entirely disappear after that. Succi, in one of his many fasts of 30 days which we investigated ... required a narcotic to allay his hunger only in the first two days; in the remaining 28 he only ingested *mineral waters*, and showed no sign of suffering': Luigi Luciani, *Human Physiology* (London: Macmillan and Co. Ltd, 1911), vol. IV, 71 (my emphasis).

59 '[U]n licor opiado con cloroformo y algún antiespasmódico, del cual conserva la comisión una cantidad para pasar a su análisis, con cuya bebida adormece la sensación del hambre en los primeros días ... Después toma alguna ligera cantidad de agua alcalina para neutralizar los ácidos del estómago ... ¿Puede bastarle esto solo para ayunar 30 días o más como hace? Nosotros creemos que no ...': Badía, 'El ayunador Succi en Barcelona', 357.

60 'But what of your strange "Elixir", distilled from African herbs, which is supposed to give you extraordinary strength? He said that: "The 'elixir' is a very simple one. It is a preparation of *laudanum*, which I take merely to soothe the stomach when pain first sets in. It somewhat resembles *chlorodyne*, although rather different in taste and far more instantaneous in its effect. I only use it as a pain-killer; it has none of the wonderful properties ascribed to it': *Pall Mall Gazette*, 26 Feb. 1890, 2 (my emphasis).

61 'The Fasting Man', *British Medical Journal*, 21 (1890), 1444.

62 'Succi Wins his Race with Starvation', *New York Herald*, 21 Dec. 1890.

63 *Cenni Biografici. Giovanni Succi. Esploratore d'Africa. Già delegato della 'Società di Commercio coll'Africa' di Milano e membro della 'Società psicologica di Madrid'* (Rome: Tipografia Economica, 1897), p. 14.

64 Many of these liquors were extracts of plants with hundreds of different organic compounds. Their systematic separation and further chemical identification posed a very hard challenge for the academic science of the late nineteenth century.

65 In spite of these controversies, Succi's explanations were in tune with medical discussions on the role of the nervous system in the feeling of hunger, the capacity to control it, and the way in which morphine and opium could calm it. 'Charles Richet a mis en évidence le rôle du système nerveux dans l'appétit en supprimant toute appétence en un chien par injection de quelques centigrammes de morphine ... Il est probable que la mort chez les fumeurs d'opium est due à l'inappétence absolue qui entraine consécutivement

l'inanition … tous les jeûneurs de profession absorbaient une mixture souvent ténue secrète, mais qu'on peut supposer aisément être une vulgaire potion de morphine': Henri Labbé, 'L'inanition. Ses aspects physiologiques et sociaux', *Revue scientifique*, 2 May 1890, 548–549.

66 John Ayto, 'Fernet Branca', *The Diner's Dictionary: Word Origins of Food and Drink*, 2nd ed. (Oxford: Oxford University Press, 2012).

67 See http://historiaycuriosidadesdelilusionismo.blogspot.com/2010/03/artistas-del-hambre-los-ayunadores.html (accessed 20 Jun. 2020).

68 *San Francisco Chronicle*, 25 May 1905, 13.

69 See https://oztypewriter.blogspot.com/2020/11/hunger-games-have-typewriter-will-fast.html (accessed 15 Apr. 2020). 'With Dr. Hommel's Haematogen, frail, anaemic, nervous people, convalescents who have been run down as a result of injuries or exertion will find an invigorating herbal remedy that has been very well received by thousands of doctors … Aktiengesellschaft Hommel's Haematogen, Zürich.' See also www.worthpoint.com/worthopedia/1915-ad-dr-hommels-haematogen-remedy-435460575 (accessed 12 Feb. 2019).

70 'Gloires des caves d'autrefois, triomphant Bordeaux, fier Bourgogne, Champagne étincelant, clos Vougeot, Moulin à vent, Pomard, Chambertin, Asti et Malaga, doux Muscat, Argenteuil grande marque, qu'allez-vous devenir? L'unique cru de Hunyadi Janos [une eau minérale] va vous remplacer': *La Caricature*, 11 Dec. 1886, 404.

71 'Even if one admitted that specific mineral principles that could be analysed chemically gave waters their therapeutic powers, chemical analysis still brought little knowledge of therapeutic effects beyond provoking speculation about how certain waters worked … Collecting data, on the model of public health statistics, seemed in fact one of the few ways to bridge the gap between chemistry and therapeutics': George Weisz, 'Water Cures and Science: The French Academy of Medicine and Mineral Waters in the Nineteenth Century', *Bulletin of the History of Medicine*, 64 (3) (1990), 406.

72 '[P]ar l'absorption d'une certaine quantité d'eaux très riches en matières salines, comme l'eau de Vichy et l'eau d'Hunyadi, M. Succi se gardait contre les accidents très graves que résultent de la privation des sels continues dans les aliments solides – des troubles profondes du system nerveux (déminéralisation). L'eau entre pour les deux tiers dans la composition de notre organisme. Elle s'élimine contentement par la peau, les muqueuses digestives et respiratoires, par le rein et diverses autres glandes. Il faut donc la remplacer incessantement, et l'homme a besoin d'en absorber en moyenne trois litres par 24 heures. Une grande partie est fournie par les aliments': quoted by Albert de Rochas, *La suspension de la vie* (Paris: Dorbon-ainé, 1913), 30.

73 'On voit des individus résister, pendant un temps fort long, à l'inanition, à l'autophagie, pourvu qu'on leur laisse boire de l'eau a discrétion. *L'eau est le milieu nutritif par excellence*, l'agent d'équilibration de l'assimilation, le pondérateur des échanges organiques qui constituent la vie et qui entretiennent l'intégrité de la santé': Monin, *L'hygiène de l'estomac*, 269 (my emphasis).

74 Douglas P. Mackaman, *Leisure Settings: Bourgeois Culture, Medicine, and the Spa in Modern France* (Chicago: University of Chicago Press, 1998); Matthew D. Eddy, 'The Sparkling Nectar of Spas; or, Mineral Water as a Medically

Commodifiable Material in the Province, 1770–1805', in Ursula Klein and Emma Spary (eds.), *Materials and Expertise in Early Modern Europe: Between Market and Laboratory* (Chicago: University of Chicago Press, 2010), 200; Christopher Hamlin, 'Chemistry, Medicine, and the Legitimization of English Spas, 1740–1840', *Medical History*, 10 (1990), 68.

75 Christopher Hamlin, *A Science of Impurity: Water Analysis in Nineteenth-Century Britain* (Bristol: Adam Hilger, 1990), 10–11.

76 Ibid., 16–47; Weisz, 'Water Cures and Science'.

77 Ignacio Suay-Matallana, 'Between Chemistry, Medicine and Leisure: Antonio Casares and the Study of Mineral Waters and Spanish Spas in the Nineteenth Century', *Annals of Science*, 73 (3) (2016), 289–302.

78 Monin, 'Succi', 775–776.

79 Luciani, *Fisiologia del digiuno*, 118.

80 Paul E. Howe, H. A. Mattill, and P. B. Hawk, 'The Influence of an Excessive Water Ingestion on a Dog after Prolonged Fast', *Journal of Biological Chemistry*, 10 (1911), 417–432.

81 Plinio Schivardi, *La meravigliosa Acqua Antilitiaca Fiuggi in Anticoli Di Campagna. Notizie storiche ricerche chimico-analitiche caratteri particolari dell'acqua sue sorprendenti virtù curative dichiarazioni scientifiche* (Rome: Editore Giuseppe Forastieri, 1898).

82 See http://historiaycuriosidadesdelilusionismo.blogspot.com/2010/03/artistas-del-hambre-los-ayunadores.html (accessed 22 May 2020).

83 See www.bottlepickers.com/bottle_articles276.htm (accessed 17 Oct. 2018).

84 Ibid.

85 Benavent, *Ayuno Succi*, 7.

86 Among them: 'le savant professeur Bouchart [et] … le docteur Ulex, chimiste expert de la ville de Hambourg … trouve son application dans les maladies de intestines, la constipation, la congestion, fièvres gastriques, et en générale dans tous les engorgements abdominaux. Cette eau purge rapidement et sans irritation gastro-intestinale: elle a l'avantage de pouvoir être administrée à petite dose, et son effet est immédiat': www.worthpoint.com/worthopedia/1900s-spain-rubinat-llorach-medicinal-1808442765 (accessed 18 Jul. 2019).

87 In Succi's fast in New York, 1890: 'Ups and Downs of Succi's Spell', *New York Herald*, 21 Dec. 1890.

88 Barbara Gronau, 'Asceticism Poses a Threat: The Enactment of Voluntary Hunger', in Alice Lagaay and Michael Lorber (eds.), *Destruction in the Performative* (Amsterdam: Rodopi, 2012), 99–109.

89 See www.showhistory.com/hungerartists.html (accessed 13 Aug. 2017).

90 'Sgr G. Succi geniesst ausschliesslich nur Becker's Sauerstoffwasser "Ozonin". [Es] ist das beliebteste und gesundeste Tafel- und Erfrischungsgetränk der Gegenwart. Einzige Spezialfabrik Carl Becker, Hannover, Hallerstrasse, 12. Fernsprecher Nr. 4465. Becker's Sauerstoffwasser "Ozonin" ist in allen messgebenden feineren Restaurants, Hôtels, Cafés und Waldwirtschaften zu haben': https://esferapublica.org/nfblog/el-hambre-de-los-artistas/ (accessed 17 Oct. 2018).

91 *La Patria*, 2 May 1960.

92 'Superiores á todas como aguas de mesa. Insuperables como aguas medicinales. Miles curados, miles curándose. Únicas en el mundo que destruyen

los cálculos hepáticos y renales y curan enfermedades del estómago … Cada botella lleva el sello oficial del Ayuntamiento de Tehuacán, como garantía de Legitimidad. Aconsejadas por los más reputados Médicos y Clínicos de la República': *El Imparcial*, 7 Aug. 1905.

93 'Puedo con toda conciencia certificar que el agua mineral "Cruz Roja" de Tehuacán es agua purísima, ligera y agradable al paladar, pudiendo competir con la mejor agua mineral extranjera, según consta en el certificado médico adjunto. Que el agua "Cruz Roja" de Tehuacán sea verdaderamente pura y ligera é inmensamente terapéutica para el estómago lo prueba y da fe mi experimento de 30 días de ayuno, que bebiendo esta agua, mi organismo no ha sentido durante todo el tiempo de la inanición ningún malestar. Debo pues decir que el agua mineral "Cruz Roja" es agua alcalina muy medicinal para la cura del estómago, y eficacísima contra los cálculos hepáticos y renales. Si recomiendo el agua "Cruz Roja" de Tehuacán, no hago más que cumplir un deber de humanidad para todos los enfermos que sufren el estómago, hígado, bazo y riñones. De usted afectísimo atento y S.S. G. Succi': ibid.

94 Ibid.

95 'El Sr Dr Fortunato Hernández. De la Escuela Nacional de Medicina de México, de la Sociedad Mexicana de Cirugía, de la Sociedad de Antropología de París, de la Sociedad Filológica Francesa, etc., recomienda las famosas aguas minerales marca "Cruz Roja" en las enfermedades del riñón, hígado, estómago e intestinos. Para evitar falsificaciones e imitaciones, cada botella lleva el sello oficial del Ayuntamiento de Tehuacán, como garantía de legitimidad … De venta en todas las droguerías, cantinas y restaurants'; 'Son transcurridos quince días desde que comencé mi ayuno. Parece que mi bienestar aumenta, conforme se verifica mi adelgazamiento … Tengo para mí, que las aguas que bebo deben de poseer virtudes inestimables, cuando he ingerido una dosis de ellas … Este día es el 26 de mi ayuno y mi sorpresa es mayor que al principio. No me cabe la menor duda, estas aguas minerales "Cruz Roja" de Tehuacán, encierran en su contenido el secreto de la salud ¿Cómo puede haber enfermos de estómago en un país que posee un manantial como el de Cruz Roja? Creo sinceramente que, si en Europa conocieran el grado de pureza de estas aguas y la feliz combinación de sus sales, no dejarían a los mexicanos un solo litro de ellas … G. SUCCI': ibid.

6 Politics

1 Charles Monselet (1825–1888) was a French journalist known as 'the king of the gastronomes'. 'Le mot de la science expérimentale est prononcé: *on peut vivre de faim*. Les *Facultés* stupéfaites s'accordent à le reconnaître … pour les *socialistes*, c'est l'extinction du paupérisme. Pour les *humanitaires* c'est l'abolition de l'anthropophagie … Pour [Charles] *Monselet* c'est la perte d'une spécialité. Pour l'*ouvrier* c'est un lundi éternel': Caliban, 'L'art de vivre de faim', *Le Figaro*, 28 Oct. 1886, front page (my emphasis).

2 Richard Altick, *The Shows of London* (Cambridge, Mass.: Belknap Press, 1978), 3.

3 Tony Bennett, *The Birth of the Museum: History, Theory, Politics* (London: Routledge, 1995), 59, 63.
4 Agustí Nieto-Galan, 'Antonio Gramsci Revisited: Historians of Science, Intellectuals, and the Struggle for Hegemony', *History of Science*, 46 (2011), 453–478.
5 Tony Bennett, 'The Exhibitionary Complex', *New Formations*, 4 (1988), 73–102; Bennett, *The Birth of the Museum*.
6 Bennett, *The Birth of the Museum*.
7 Agustí Nieto-Galan, *Science in the Public Sphere: A History of Lay Knowledge and Expertise* (London: Routledge, 2016).
8 Emma Spary and Anya Zilberstein, 'On the Virtues of Historical Entomophagy', *Osiris*, 35 (2020), 15.
9 'M. Succi travaille donc, sans y penser peut-être, à résoudre la question sociale au profit du pauvre contre le riche': 'Chronique', *La semaine des familles*, 2 Oct. 1886, 431–432.
10 Bryan S. Turner, 'The Government of the Body: Medical Regimens and the Rationalization of Diet', *British Journal of Sociology*, 33 (2) (1982), 254; James Vernon, *Hunger: A Modern History* (Cambridge, Mass.: Belknap Press, 2007).
11 '[D]ans la population parisienne, les déchéances, les maladies, la tuberculose …, étaient fonction d'une alimentation déplorable … *La question de l'inanition, élargie à celle de la nourriture insuffisante, déborde singulièrement le cadre physiologique qu'on se plait d'abord à lui tracer. Il s'agit d'une question sociale*, d'une question de vitalité de d'avenir de la race qui ne pourrait être résolue que par l'instruction et l'éducation progressive de tous': Henri Labbé, 'L'inanition. Ses aspects physiologiques et sociaux', *Revue scientifique*, 2 May 1890, 551 (my emphasis). See also Henri Labbé, *Principes de la diethétique moderne* (Paris: Bailliere et Fils, 1904). Labbé mentioned the work of Dr Landouzy 'sur l'alimentation des parisiens' and the description of tuberculosis as the 'maladie de misère'.
12 Rex Taylor and Annelie Rieger, 'Rudolf Virchow on the Typhus Epidemic in Upper Silesia: An Introduction and Translation', *Sociology of Health and Illness*, 6 (2) (1984), 201–217.
13 Alfons Labisch, *Homo Hygienicus: Gesundheit und Medizin in der Neuzeit* (Frankfurt: Campus Verlag, 1992); Avner Offer, 'Body Weight and Self Control in the US and Britain since the 1950s', *Social History of Medicine*, 14 (1) (2001), 79–106.
14 Sigal Gooldin, 'Fasting Women, Living Skeletons and Hunger Artists: Spectacles of Body and Miracles at the Turn of a Century', *Body and Society*, 9 (2) (2003), 47.
15 'George Francis Train Starving Himself to Death', *Dundee Courier*, 7 Jan. 1878, 4.
16 Robert A. Gunn, *40 Days without Food: A Biography of Henry S. Tanner MD, Including a Complete and Accurate History of His Wonderful Fast, 42 Days in Minneapolis, and 40 Days in New York City, with Valuable Deductions* (New York: Albert Metz, 1880), preface.
17 'L'homme mange beaucoup plus qu'il devrait le faire, et c'est là … une des causes principales de la plupart de ses maladies, plus communes chez lui

que chez les animaux': Henri Perrussel, 'La diète prolongée volontaire', *Les Annales politiques et littéraires*, 10 Mar. 1886, 220–221.

18 Ibid.

19 'Il est certain que nos repas sont trop rapprochés, soit par le fait des usages admis, soit par sensualité, soit par une sorte d'idiosyncrasie acquise, qui ne nous permet pas de résister à la faim. Il serait plus simple, comme le plus naturel, *de ne faire que deux repas par jour*, celui du matin et celui du soir (entre sept et huit heures). Voilà quel serait le problème social à résoudre et qui, du moins, serait d'une utilité pratique, tandis que nos jeûneurs ne servent qu'à nourrir la curiosité des badauds': Adolphe Burggraeve, *Longévité humaine per la médecine dosimétrique ou la médecine dosimétrique à la portée de tout le monde, avec ses applications à nos races domestiques* (Paris: Librairie et gare de chemin de fer, 1887), 195–196 (my emphasis).

20 Ibid., ch. XIV, 'Les jeûneurs', 193–203.

21 'Se tutti potessero abituarsi a magiar poco ed una volta al giorno si avrebbe maggior tempo per coltivare la mente': Giovanni Chiverny, *Del Signor Succi e del suo digiuno* (Milan: Tipografia di Giacomo Pirola, 1886), 37.

22 'Empty Bread Baskets as a Steady Diet', *New York Herald*, 9 Nov. 1890 (my emphasis).

23 Louis Landuzy, *Aperçus de médecine sociale. Extrait de la Revue de Médecine* (Paris: Félix Alcun, 1905), 24.

24 'Recuerdan nuestros lectores los famosos ayunadores Succí y Merlatti? Pues ahora en París van a tener no un ayunador, sino una ayunadora, una artista americana, Miss Nelson. Como sus predecesores, se rodeará de una junta de médicos que la vigilarán noche y día, y se asegurarán de la verdad de sus declaraciones. Miss Nelson se obliga a no tomar diariamente más que un vaso de una bebida especial. *La pretensión de Miss Nelson es la de dar a los pobres el medio de pasarse sin comer y sin comprometer su salud.* Empezará sus experimentas hoy domingo en el Gran Hotel de París, y dará, en el transcurso de su ayuno, cuatro conciertos a beneficio de los pobres de la capital francesa': *El Heraldo de Madrid*, 24 Jan. 1892 (my emphasis).

25 'La cuisine a tué plus d'hommes que la guerre ... ce n'est pas une ironie de la Providence que ces expériences de Succi, Merlatti et autres qui viennent démontrer aux hommes de notre époque qui n'aspirent qu'à la jouissance sensuelle, que le corps peut être soumis, sans trop y perdre, à un jeûne répété, soutenu, abusive même; que cette société des végétariens qui vont prouver – en le pratiquant ouvertement – que la nourriture végétale peut faire vivre l'homme aussi bien et plus normalement même que l'alimentation carnivore': L'abbé Vallée, curé de Monts, 'Essai d'explication théorique et scientifique du jeûne extraordinaire des italiens Succi et Merlatti', *Annales de la Société d'agriculture, sciences, arts et belles lettres d'Indre-et-Loire*, 67 (1887), 36.

26 Leon Tolstoi, 'Notre alimentation', *Revue Scientifique*, 20 Aug. 1892, 235.

27 '[L]a première condition d'une vie morale est l'abstinence, la première condition de l'abstinence est le jeûne': ibid., 230.

28 Charles Richet, 'L'alimentation et le luxe. Reponse à L. Tolstoi', *Revue scientifique*, 50 (1892), 385–391.

29 Ibid., 389.

30 '[I]l peut y avoir pour l'homme quelque avantage à manger de la viande, quoique la viande ne soit indispensable. C'est l'opinion adoptée, avec raison, croyons-nous, par la majorité des physiologistes': ibid., 391.
31 Hans-Christian Gunga and Karl A. Kirsch, 'Nathan Zuntz (1847–1920), a German Pioneer in High Altitude Physiology and Aviation Medicine, Part I: Biography', *Aviat Space Environmental Medicine*, 66 (2) (1995), 168–171.
32 Charles Richet, *Physiologie. Travaux du laboratoire de M. Charles Richet. Chimie physiologique. Toxicologie* (Paris: Felix Alcun, 1893), Chapter XXIX, 'L'inanition', 267–325.
33 John S. Haller, *Medical Protestants: The Eclectics in American Medicine, 1825–1939* (Carbondale: Southern Illinois University Press, 1994).
34 Linda B. Hazzard, *Scientific Fasting: The Ancient and Modern Key to Health* (New York: Grant Publications, 1927).
35 '[U]ne science de fonctionnaires et des fonctionnaires d'une seule et même administration, où l'émulation, la noble rivalité, la libre concurrence sont comme l'indépendance aussi inconnues et impossibles qu'en cage les grands coups d'aile. Voilà le mal dont on pourrait mourir. Nous allons à une bureaucratie scientifique': Victor Meunier, *Scènes et types du monde savant* (Paris: Octave Doin, 1889), 5–6.
36 Catherine Glaser, 'Journalisme et critique scientifiques. L'exemple de Victor Meunier', *Romantisme*, 65 (1989), 27–48; Agustí Nieto-Galan, 'Useful Charlatans: The Fasting Contest of Giovanni Succi and Stefano Merlatti in Paris, 1886', *Science in Context,* 33 (4) (2020), 405–422.
37 Isabelle Cavé, *Les médecins-législateurs et le mouvement hygiéniste sous la Troisième République, 1870–1914* (Paris: l'Harmattan, 2014); John Woodward and David Richards, *Health Care and Popular Medicine in Nineteenth-Century England: Essays in the Social History of Medicine* (New York: Holmes & Meier, 1977). Among Monin's titles it is worth mentioning: *Hygiène et médecine journalière*; *La lutte pour la santé*; *Les maladies épidémiques*; *L'hygiène du travail*; *La santé par l'exercice*; *Formulaire de médecine*; *La santé des riches*; *Précis d'hygiène* (with Dr Dubousquet); *Les remèdes que guérissent*; *Obésité et maigreur*; *Essai d'hygiène pratique*; *L'hygiène de l'estomac*; *Guide pratique de l'alimentation*; *Essai d'hydrologie clinique*; and *Hygiène et traitement curative des troubles digestifs.*
38 Dr. E. Gallamand (de St. Mandé), 'Le docteur Monin', *Le progrès médical,* 5 Jan. 1929, 6.
39 'Les pauvres meurent deux fois plus que les riches: la misère est la grande pourvoyeuse de la mort. Les citadins meurent deux fois plus, également, que les campagnards, précisément parce que la misère est plus fréquente dans les grandes villes': Ernest Monin, *Le propos du docteur. Médecine sociale, hygiène génerale à l'usage des gens du monde* (Paris: E. Giraud et Cie, 1885), 183.
40 '[O]bserver un régime sobre, en se rappelant que la gueule, comme l'a dit Brantôme, a fait plus de victimes que la glaive': ibid., 184. Monin was referring to the adventurer Pierre de Bourdeille (1540–1614).
41 'Léese una comunicación del Sr Succi, pidiendo al Congreso nombre una comisión para comprobar y fiscalizar su ayuno de 30 días y los ejercicios corporales (entre paréntesis, el ayunador da argumentos a los "burgueses" satisfechos para desestimar las reclamaciones de las clases obreras miserables;

pues les dirán que se puede vivir y trabajar alimentándose poco, ejemplo los Succi)': 'Congreso de Ciencias Médicas en Barcelona con motivo de su Exposición Universal. Sesión inaugural', *La Exposición*, 65 (1888), 7.

42 Enrique Sepúlveda, *La vida en Madrid* (Madrid: Ricardo Fe, 1888), 546.

43 Mr Simon, Brussels, 1886, www.sideshowworld.com/13-TGOD/2014/Hunger/Artists.html (accessed 25 May 2020).

44 *La Vanguardia*, 17 Nov. 1888, front page.

45 '[A] los fisiologistas toca determinar los efectos de la inanición y descubrirlos con su espantosa realidad, toda vez que el escalpelo es insensible a la compasión; con todo, la *caridad* nunca pierde sus derechos, y a ella se corresponde aliviar los sufrimientos humanos ... Esta debe ser la conclusión de nuestro estudio; *multipliquemos las sociedades de beneficencia,* demos más, demos siempre para ayudar a los desgraciados en su lucha por la existencia y hagamos lo imposible para que la historia no registre en adelante los dramas horribles que desgraciadamente anota en sus páginas con harta frecuencia': Máximo de Nausouty, 'La inanición', *Diario de Barcelona*, 4 Sep. 1895, 10206.

46 Carl von Voit revisited Liebig's theory on the most efficient combination of nitrogenous and non-nitrogenous foods (fats and carbohydrates) to consume with which a given animal could maintain its life. See Russell Chittenden, *Physiological Economy in Nutrition with Special Reference to the Minimal Proteid Requirement of the Healthy Man: An Experimental Study* (London: Heinemann, 1905).

47 '[C]elle qui est nécessaire pour que le poids de l'être vivant n'augmente ni ne diminue, mais conserve la valeur exacte qu'il possède, à moment où il ne serait ni trop maigre ni trop gras': Wilfrid de Fonvielle, *Mort de faim. Étude sur les nouveaux jeûneurs* (Paris: Librairie illustrée, 1886), 110.

48 'L'exercice quotidienne que se donnent les jeûneurs, quand ils vont faire leur promenade au Bois de Boulogne, dans les rues de Turin, ou autour du lac de Chicago, ne peuvent s'accomplir qu'au détriment du capital viande et graisse, que la combustion pulmonaire consommera': ibid., 112.

49 'L'équivalent mécanique de la chaleur nous apprend que l'homme qui exerce une action musculaire doit consommer une quantité de carbone plus grand que celui qui ne fait rien': ibid., 112–113.

50 Hans-Christian Gunga and Karl A. Kirsch, 'Nathan Zuntz (1847–1920) – a German Pioneer in High Altitude Physiology and Aviation Medicine, Part II: Scientific Work', *Aviat Space Environmental Medicine*, 66 (2) (1995), 172–176. Zuntz invented the Gasuhr (dry gas measuring device).

51 Angelo Mosso, *Fatigue* (London: Swan Sonnenschein & Co. Ltd, 1904), 225 (Italian original ed., 1891).

52 Harmke Kamminga and Andrew Cunningham (eds.), *The Science and Culture of Nutrition, 1840–1940* (Amsterdam: Rodopi, 1995), 7. See also 'The Virtual Laboratory: Essays and Resources on the Experimentalization of Life', http://vlp.mpiwg-berlin.mpg.de/index_html (accessed 15 Feb. 2019).

53 Richard Gilliespie, 'Industrial Fatigue and the Discipline of Physiology', in Gerald Geison (ed.), *Physiology in the American Context, 1880–1940* (Bethesda, Md.: American Physiological Society, 1987), 238.

54 Daniel Rees, *Hunger and Modern Writing: Melville, Kafka, Hamsun and Wright* (Cologne: Map, 2016).

55 Elizabeth Neswald and David F. Smith, *Setting Nutritional Standards: Theory, Policies, Practices* (Rochester, N.Y.: Rochester Studies in Medical History, 2017), 12.
56 Corinna Treitel, 'Nutritional Modernity: The German Case', *Osiris*, 35 (1) (2020), 183–203.
57 Carol Harrison and Ann Johnson, 'Science and National Identity', *Osiris*, 24 (2009), 1–14.
58 Paul Greenhalgh, *Ephemeral Vistas: The Expositions Universelles, Great Exhibitions and World's Fairs, 1851–1939* (Manchester: Manchester University Press, 1988).
59 Fonvielle, *Mort de faim*, 279–80.
60 'Ces jeûnes font partir d'un complot contre la dignité humaine et l'honneur de la Patrie. Quelle ne serait pas, en effet, la joie des états monarchiques si des absurdités … étaient reconnues et proclamées par les pouvoirs publiques, si les prétentions qui ont osé se produire à l'Académie de médecine, et à l'Académie des sciences morales étaient acceptés par le Sénat et la Chambre des députés d'une république qui s'enorgueillit de ne baser son état social et ses politiques que sur la raison': ibid., 280.
61 Jérôme Auvinet, *Charles-Ange Laisant. Itinéraires et engagements d'un mathématicien de la Troisième République* (Paris: Hermann, 2013).
62 'La experiencia es curiosa y rara, lo sería más en otro país que no fuera el nuestro, porque aquí donde una gran parte de la población vive del aire … aquí donde los albañiles y carpinteros que trabajan en las construcciones tienen bastante con un pimiento o un pedazo de queso; aquí donde los maestros de escuela no comen, al menos que yo sepa; no deben poder comer, mejor dicho, por las horribles intermitencias que sufren para cobrar sus haberes; aquí, repito, el ayuno Succi no sorprende todo lo que debiera': Sepúlveda, *La vida en Madrid*, 546.
63 'Porque si es así, Succi sería un verdadero Salvador de la humanidad, y habría resuelto un problema más arduo y más difícil que el de vivir sin trabajo, cuál sería el de vivir sin comer … en esta obscuridad, en esta duda no aclarada, estriba el poco interés que el espectáculo ha despertado en Madrid, y la escasez de visitantes en la sala del Teatro Felipe': ibid., 547.
64 'Confesad que os habéis equivocado; durante vuestro ayuno en el Palacio de Ciencias, habéis sido un símbolo, una verdadera abstracción. Esta abstinencia prolongada de treinta días en medio de las obras de enseñanza, de los productos de asociaciones científicas y literarias, de la muestra de los esfuerzos de la ciencia y del estudio en España, simboliza las privaciones y los ayunos á que viene condenado el infeliz que se dedica a especulaciones científicas y a la enseñanza, desdeñando la tauromaquia, la política y la Bolsa. Muchas veces parece que el azar tenga conciencia, y nunca mayor motivo tuvimos para sospecharlo, que en el hecho de haberse realizado vuestro ayuno en el Palacio de Ciencias': Federico Rahola, 'Carta a Succi', *La Vanguardia*, 15 Oct. 1888, 2.
65 Agustí Nieto-Galan, 'The Images of Science in Modern Spain: Rethinking the "Polémica"', in Kostas Gavroglu (ed.), *The Sciences in the European Periphery during the Enlightenment* (Dordrecht: Kluwer, 1998), 65–86.
66 Corinna Treitel, *Eating Nature in Modern Germany: Food, Agriculture and Environment, c. 1870 to 2000* (Cambridge: Cambridge University Press, 2017).

67 Ibid.
68 Treitel, 'Nutritional Modernity'.
69 'The Famous Faster: An Interview with Signor Succi', *Pall Mall Gazette*, 26 Feb. 1890, 2.
70 Spary and Zilberstein, 'On the Virtues of Historical Entomophagy'.
71 Jules Demolliens, 'Tout au jeûne', *La Caricature*, 20 Nov. 1888, 378–379.
72 Bennett, *The Birth of the Museum*.
73 Turner, 'The Government of the Body', 258.
74 Margaret Cohen, 'Walter Benjamin's Phantasmagoria', *New German Critique*, 48 (1989), 87–107. I am indebted to Robert Caner-Liese for introducing me to Benjamin's work and world view.
75 H. Hazel Hahn, *Scenes of Parisian Modernity: Culture and Consumption in the Nineteenth Century* (London: Palgrave Macmillan, 2009), 1–14.
76 Cohen, 'Walter Benjamin's Phantasmagoria'.
77 Georg Franck, 'The Economy of Attention', *Journal of Sociology*, 558 (1) (2019), 8–19.
78 '[S]ituation des individus qui se sont trouvés pris, par exemple, sous des éboulements ... les malheureux ainsi séparés du reste du monde savent qu'on ne peut pas arriver à leur secours: qu'il faut, pour parvenir jusqu'à eux, percer des galléries, déblayer quelques centaines de mètres cubiques de terre et de pierres. Dans ces conditions, la privation de nourriture est parfois fort longue ... Les mineurs de Bos-Mouzil sont restés huit jours enfermés à la suite d'un éboulement, sans souffrir beaucoup. Ces derniers ont d'ailleurs donné un exemple bien rare de solidarité et de charité humaines. En général, les individus soumis aux angoisses de la faim montrent un égoïsme féroce ... On cite dans ces jeûnes forcés, d'autres exemples, non plus de mineurs, mais de naufrages': Richet, 'L'inanition', 306–7.
79 Ibid., 301.
80 'Chicago, July 12. Griscom's long fast concluded at noon. In the 45 days he has lost just 50 pounds, his weight today being 147.5 pounds, temperature 98, pulse 55. About 200 people were present to see him eat his first meal. Griscom stated to the audience that he had procured everything his fancy suggested, not because he expected to eat, but to satisfy his imagination, and have just what he might want. He thought he could eat a full meal without injury, but he did not intend to do so. He believes that there was a great virtue in fasting; he was willing to fast longer than necessary to show that this was a good remedial agent. It had once cured him when the physicians had said that he must die. He did not believe in indiscriminate or unnecessary prolonged abstemiousness, but in moderate and well-considered fasting. At the conclusion of his speech, he took a goblet of milk and with the same deliberation that had marked his motions all the morning commenced his meal. Chicago July 13. The interest in John Griscom, the faster, has not yet subsided. He visited the Grand Opera House last night and felt better this morning for having seen the play. For breakfast he ate two slices of molasses cake and drank a glass of cream. He tried to swallow some tea but did not relish it. He is still quite weak with the unpleasant sensations in the bed, but is mending; unlike Tanner he has no exaggerated appetite, but eats less and desires less than an ordinary man. *He insists that his fast has*

been of incalculable value to science': *Weekly Wisconsin*, 20 Jul. 1881, 7 (my emphasis).

81 'Record Is Set by M'Swiney', *Washington Herald*, 14 Oct. 1920, 1.

Conclusion

1 Robert Alexander Gunn, *40 Days without Food: A Biography of Henry S. Tanner MD, Including a Complete and Accurate History of His Wonderful Fast, 42 Days in Minneapolis, and 40 Days in New York City, with Valuable Deductions* (New York: Albert Metz, 1880), Chapter VI, 'Deductions', 90–94.

2 Ibid., 90 (my emphasis).

3 Physiology textbooks usually refer to Charles Chossat's famous text: *Recherches expérimentales sur l'inanition. Mémoire auquel l'Académie des Sciences a décerné en 1841 le prix de physiologie expérimentale* (Paris: Imprimérie royale, 1843). See Jean Jacques Dreifuss, 'Charles Chossat (1796–1875), physiologiste, médecin et homme politique genevois', *Gesnerus*, 45 (1988), 239–261.

4 Gunn, *40 Days without Food*, 90.

5 Ibid., 91.

6 Ibid., 90–91.

7 Ibid., 92.

8 Ibid.

9 Ibid. (my emphasis).

10 Ibid.

11 Ibid., 94.

12 'The Fasting Man', *British Medical Journal*, 21 (1880), 1446 (my emphasis).

13 George M. Gould and Walter L. Pyle, *Anomalies and Curiosities of Medicine* (Philadelphia: W. B. Saunders, 1897), 413–414 (my emphasis).

14 *British Medical Journal*, 21 Jun. 1890, 1446.

15 Dr. Luis C. Maglioni, *Mis 37 días de ayuno (auto-experiencia). A propósito de los casos de Mr MacSwiney y sus 11 compañeros, ayunadores políticos en las prisiones de Brixton y de Cork. 'Huelga de Hambre'* (Buenos Aires: Imprenta San Martín, 1920).

16 Luis C. Maglioni, *L'oeil de Marconi* (London: Spottiswoode, 1913); Luis C. Maglioni, *Arte y ciencia. A propósito de la conferencia dada por el Dr Antonio Caso (embajador mejicano) en el Consejo Nacional de Mujeres el 27 de septiembre de 1921* (Buenos Aires: San Martín, 1922); Luis C. Maglioni, *Cirujía* (Buenos Aires: San Martín, 1923).

17 Edward H. Dewey, *The No-breakfast Plan and the Fasting-cure* (Meadville, Pa.: published by the author, 1900); Marie Griffith, 'Apostles of Abstinence: Fasting and Masculinity during the Progressive Era', *American Quarterly*, 52 (4) (2000), 599–638.

18 Francis J. Costello, *Enduring the Most: The Biography of Terence MacSwiney* (Dingle, Ireland: Brandon Books, 1996).

19 On hunger strikes and force-feeding, see Ian Miller, *A Modern History of the Stomach: Gastric Illness, Medicine and British Society, 1800–1950* (London: Pickering & Chatto, 2011); Ian Miller, 'A Prostitution of the Profession?

Forcible Feeding, Prison Doctors and the British State, 1909–1914', *Social History of Medicine*, 26 (2) (2013), 225–245; Melinda Grimsley-Smith, 'Revisiting a "Demographic Freak": Irish Asylums and Hidden Hunger', *Social History of Medicine*, 25 (2) (2011), 307–323; Turner, 'The Government of the Body'.

20 Jennian F. Geddes, 'Culpable Complicity: The Medical Profession and the Forcible Feeding of Suffragettes, 1909–1914', *Women's History Review*, 17 (1) (2008), 79–94.

21 Maud Ellmann, *The Hunger Artists: Starving, Writing and Imprisonment* (London: Virago Press, 1993), 17.

22 Margaret Cohen, 'Walter Benjamin's Phantasmagoria', *New German Critique*, 48 (1989), 87–107.

23 Mark Mazower, *Dark Continent: Europe's Twentieth Century* (London: Allen Lane, 1998).

24 Federico Rahola, 'Carta a Succi', *La Vanguardia*, 15 Oct. 1888, 2.

25 'Therapeutic starvation was finally stigmatized as an unsafe procedure exposing the patient to an undue risk of physical danger': Jean Hervé Lignot and Yvon LeMaho, 'A History of Modern Research into Fasting, Starvation and Inanition', in Marshall D. McCue (ed.), *Comparative Physiology of Fasting, Starvation, and Food Limitation* (Dordrecht: Springer, 2012), 10. There were some exceptions such as Arnold Ehret, *Rational Fasting for Physical, Mental and Spiritual Rejuvenation* (Los Angeles: Health Center of Los Angeles, 1926). I am indebted to A. Mulberger and Andrea Graus for this publication. Arnold Ehret (1866–1922) was a famous German naturopath who promoted therapeutic dieting and fasting from a heterodox, charlatan position.

26 See, for example, the case of the Buchinger Wilhelmi therapeutic fasting: www.buchinger-wilhelmi.com/wp-content/uploads/2020/08/BW_Pressemappe_Online_EN.pdf (accessed 15 Feb. 2021).

Bibliography

Archives and Libraries

American Philosophical Society Library, Philadelphia
Biblioteca de Catalunya, Barcelona
Biblioteca de la Reial Acadèmia de Ciències i Arts de Barcelona
Biblioteca de l'Ateneu Barcelonès, Barcelona
Biblioteca Marucelliana, Florence
Biblioteca Nacional de España, Madrid
Biblioteca nazionale centrale di Firenze, Florence
Biblioteca virtual de prensa histórica, Madrid
Bibliothèque nationale de France, Paris
Bodleian Libraries, University of Oxford
British Library, London
British Newspaper Archive, London
Carnegie Institution of Washington Archive, Washington, D.C.
Gallica, Paris
Gilbert and Sullivan Archive, online
Hemeroteca Arxiu Històric de la Ciutat, Barcelona
Hemeroteca Nacional Digital de México, Mexico City
Internet Archive, San Francisco
John Johnson Collection of Printed Ephemera (Bodleian Libraries), University of Oxford
Mary Evans Picture Library, London
National Fairground and Circus Archive, Sheffield, UK
Österriechischen Nationalbibliothek, Vienna
Reuters Archive, London
Wellcome Collection, London

Newspapers and Periodicals

American Journal of Physiology
American Journal of the Medical Sciences
Annales de la Société d'agriculture, sciences, arts et belles lettres d'Indre-et-Loire

Annales politiques et littéraires, Les
Archiv für pathologische Anatomie und Physiologie und für klinische Medicin
Archives générales de médicine
Belfast Telegraph (UK)
Bolletino della Sociéta geogràfica italiana
British Medical Journal
Bulletin de la Société de géographie de l'Ain
Bulletin de la Société scientifique d'hygiène alimentaire et d'alimentation rationnelle de l'homme
Bulletin du progrès médical
Caricature, La
Chattanooga Daily Times
Chemist and Druggist
Chicago Herald
Clinique, La. Organe de l'Homéopathie complexe
Columbus Daily Enquirer (Ohio)
Consultor homeopático, El
Correspondencia de España, La
Corriere Spiritico, Il
Criterio Espiritista, El
Decatur Herald (Georgia)
Diario de Barcelona
Diario del hogar, El (Mexico City)
Diplomate, Le
Dundee Courier (UK)
Écho du merveilleux, L'
Época, La
Exposición, La
Figaro, Le
Framlingham Weekly News (UK)
Gazette du village, La
Gazzetta Ufficiale del Regno d'Italia
Globe (UK)
Heraldo de Madrid, El
Huntly Express (UK)
Illustrated London News
Illustrated Police News
Illustration européenne, L'
Ilustración Española y Americana, La
Imparcial, El
Impartial, L'
Journal de physiologie et de pathologie générale
Journal des connaissances médicales
Lancet, The
Loro, El (Cádiz)
Manchester Courier (UK)
Morning Post (UK)
Mundo, El

Nature, La
New York Daily Tribune
New York Herald
New York Times
Notre Dame Scholastic
País, El
Pall Mall Gazette
Patria, La
People, The (UK)
Petit journal, Le
Petit parisienne, Le
Poverty Bay Herald (New Zealand)
Presse, La
Proceedings of the Royal Society of Edinburgh
Progrès médical, Le
Rappel
Register (Adelaide)
Revue britannique
Revue des journaux et des livres, La
Revue scientifique
San Francisco Chronicle
Semaine des familles, La
Semaine médicale, La
Sheffield Daily Telegraph
Siglo médico, El
Skadinavian Archiv für Physiologie
Sperimentale, Lo
Stampa, La
Suffolk and Essex Free Press (UK)
Tiempo, El
Toronto Daily Mail
Tribune (New York)
United States Experiment Station Record
Univers illustré, L'
Universal de México, El
Vanguardia, La
Victoria Daily Colonist (Canada)
Vie posthume, La
Voltaire
Washington Herald
Washington Post
Weekly Wisconsin
Wiener Caricaturen
Yorkshire Post and Leeds Intelligencer
Zeitschrift für experimentelle Pathologie und Therapie
Zeitschrift für physiologische Chemie

Books and Articles

Abbatista, Guido, *Umanità in mostra. Exposizioni etniche e invenzione esotiche in Italia (1880–1940)*. Trieste: Edizioni Università di Trieste, 2013.

Abend, Lisa, 'Specters of the Secular: Spiritism in Nineteenth-Century Spain', *European History Quarterly*, 31 (4) (2004): 507–534.

Agnew, Jeremy, *Entertainment in the Old West: Theater, Music, Circuses, Medicine Shows, Prize Fighting and Other Popular Amusements*. Jefferson, N.C.: McFarland & Company, 2011.

Alkon, Paul K., *Science Fiction before 1990: Imagination Discovers Technology*. New York: Twayne, 1994.

Altick, Richard, *The Shows of London*. Cambridge, Mass.: Belknap Press, 1978.

Altman, Laurence, *Who Goes First? The Story of Self-Experiments in Medicine*. Berkeley: University of California Press, 1998.

Ameke, Wilhelm, *History of Homeopathy: Its Origin, Its Conflicts. With an Appendix of the Present State of University Medicine*. London: E. Gould and Son, 1885.

Anderson, Ann, *Snake Oil, Hustlers and Hambones: The American Medicine Show*. Jefferson, N.C.: McFarland, 2000.

Anderson, Stuart, 'Travelers, Patent Medicines, and Pharmacopoeias: American Pharmacy and British India, 1857 to 1931', *Pharmacy in History*, 58 (2016): 63–82.

Atkins, Peter J., Peter Lummel, and Derek J. Oddy (eds.), *Food and the City in Europe since 1800*. Aldershot and Burlington: Ashgate, 2007.

Atwater, Wilbur O., and Francis G. Benedict, *Experiments on the Metabolism of Matter and Energy in the Human Body, 1900–1902*. Washington, D.C.: Government Printing Office, 1903.

Baldwin, James Mark (ed.), *Dictionary of Philosophy and Psychology Including Many of the Principal Conceptions of Ethics, Logics, Aesthetics, Philosophy of Religion, Mental Pathology, Anthropology, Biology, Neurology, Physiology, Economics, Political and Social Philosophy, Philology, Physical Science, and Education, and Giving a Terminology in English, French, German and Italian*, 3 vols. New York: Macmillan, 1905.

Barrucand, Dominique, 'Freud et Bernheim', *Histoire des sciences médicales*, 20 (1986): 157–170.

Bayly, Christopher, *The Birth of the Modern World 1780–1914: Global Connections and Comparisons*. Oxford: Blackwell, 2004.

Benavent, Javier de, *Ayuno Succi*. Barcelona: Tipo-Litografía de Busquets y Vidal, 1890.

Benedict, Francis Gano, 'Chemical and Physiological Studies of a Man Fasting Thirty-One Days', *Proceedings of the National Academy of Sciences of the United States of America*, 1 (1915): 228–231.

'Effet physiologiques d'une réduction prolongée du régime alimentaire expérimentée sur 25 sujets', *Bulletin de la société scientifique d'hygiène alimentaire et d'alimentation rationnelle de l'homme*, 6 (1918): 422–430.

The Influence of Inanition on Metabolism. Washington, D.C.: Carnegie Institution of Washington, 1907.

A Study of Prolonged Fasting. Washington, D.C.: Carnegie Institution of Washington, 1915.

Bennett, Tony, *The Birth of the Museum: History, Theory, Politics*. London and New York: Routledge, 1995.

'The Exhibitionary Complex', *New Formations*, 4 (1988): 73–102.

Bensaude-Vincent, Bernadette, and Christine Blondel (eds.), *Des savants face à l'occulte, 1870–1940*. Paris: La Découverte, 2002.

(eds.), *Science and Spectacle in the European Enlightenment*. Aldershot: Ashgate, 2008.

Bernheim, Hippolyte, *De la suggestion et de ses applications à la thérapeutique*, 2nd ed. Paris: O. Doin, 1888.

'Le jeûne de Succi', *Journal des connaissances médicales*, 54 (1886): 341–343.

Blanchard, Pascal, Gilles Boetsch, and Nanette Jacomijn Snoep, *Human Zoos: The Invention of the Savage*. Arles: Actes Sud, 2011.

Bogdan, Robert, *Freak Show: Presenting Human Oddities for Amusement and Profit*. Chicago: University of Chicago Press, 1988.

Bompiani, Sofia, *Italian Explorers in Africa*. London: Religious Tract Society, 1891.

Bonah, Christian, and Anne Rasmussen, *Histoire et médicament aux XIXe et XXe siècles*. Paris: Editions Glyphe, 2005.

Bondeson, Jan, *The Lion Boy and Other Medical Curiosities*. Stroud: Amberley Publishing, 2018.

Borell, Merriley, 'Instrumentation and the Rise of Modern Physiology', *Science and Technology Studies*, 5 (2) (1987): 53–62.

'Instruments and an Independent Physiology: The Harvard Physiological Laboratory, 1871–1906', in Gerald Geison (ed.), *Physiology in the American Context, 1850–1940*. Bethesda, Md.: American Physiological Society, 1987, 239–321.

Brock, Pope, *Charlatan: The Fraudulent Life of John Brinkley*. London: Weidenfeld & Nicolson, 2008.

Brock, William H., *Justus von Liebig: The Chemical Gatekeeper*. Cambridge: Cambridge University Press, 2002.

Brockliss, Laurence W. B., and Colin Jones, *The Medical World of Early Modern France*. Oxford: Clarendon, 1997.

Brozek, Josef, 'Six Recent Additions to the History of Physiology in the USSR', *Journal of the History of Biology*, 6 (2) (1973): 317–334.

Brumberg, Joan Jacobs, *Fasting Girls: The Emergence of Anorexia Nervosa as a Modern Disease*. Cambridge, Mass.: Harvard University Press, 1988.

Brun, Philippe, *Albert Robida. Fantastique et science-fiction*. Paris: P. Horay, 1980.

Burggraeve, Adolphe, *Longévité humaine par la médecine dosimétrique ou la médecine dosimétrique à la portée de tout le monde, avec ses applications à nos races domestiques*. Paris: Librairie et gare de chemin de fer, 1887.

Busby, Siân, *A Wonderful Little Girl: The True Story of Sarah Jacob, the Welsh Fasting Girl*. London: Short Books, 2003.

Butler, Stella V. F., 'Center and Peripheries: The Development of British Physiology, 1870–1914', *Journal of the History of Biology*, 21 (1988): 473–500.

Bycroft, Michael, 'Iatrochemistry and the Evaluation of Mineral Waters in France, 1600–1750', *Bulletin of the History of Medicine*, 91 (2) (2017): 303–330.

Bynum, William, *Science and the Practice of Medicine in the Nineteenth Century*. Cambridge: Cambridge University Press, 1994.

Bynum, William, and Roy Porter, *Medical Fringe and Medical Orthodoxy, 1750–1850*. London: Croom Helm, 1987.

Cahan, David, *Helmholtz: A Life in Science*. Chicago: University of Chicago Press, 2018.

Canadelli, Elena, 'Scientific Peep Show: The Human Body in Contemporary Science Museums', *Nuncius*, 26 (2011): 159–184.

Carlson, Anton Julius, *The Control of Hunger in Health and Disease*. Chicago: University of Chicago Press, 1916.

Carpenter, Kenneth J., 'A Short History of Nutritional Science: Part 1 (1785–1885)', *Journal of Nutrition*, 133 (3) (2003): 638–645.

'A Short History of Nutritional Science: Part 2 (1885–1912)', *Journal of Nutrition*, 133 (4) (2003): 975–984.

Cassata, Francesco, and Claudio Pogliano (eds.), *Scienze e cultura dell'Italia unita*. Turin: Einaudi, 2011.

Cathcart, Edward P., *The Physiology of Protein Metabolism*. London: Longmans, Green and Co., 1912.

Cenni Biografici. Giovanni Succi. Esploratore d'Africa. Già delegato della 'Società di Commercio coll'Africa' di Milano e membro della 'Società psicologica di Madrid'. Rome: Tipografia Economica, 1897.

Chadarevian, Soraya de, 'Graphical Method and Discipline: Self-Recording Instruments in Nineteenth-Century Physiology', *Studies in History and Philosophy of Science*, 24 (2) (1993): 267–291.

Chapin, Robert C., *Standard of Living among Workingmen's Families in New York City*. New York: Charities Publication Committee, 1909.

Chittenden, Russell, *Physiological Economy in Nutrition, with Special Reference to the Minimal Proteid Requirement of the Healthy Man: An Experimental Study*. London: Heinemann, 1905.

Chiverny, Giovanni, *Del Signor Succi e del suo digiuno*. Milan: Tipografia di Giacomo Pirola, 1886.

Nozioni technico-practiche per uso degli inernieri di manicomia. Milan: Pirola, 1878.

Chossat, Charles, *Recherches expérimentales sur l'inanition. Mémoire auquel l'Académie des Sciences a décerné en 1841 le prix de physiologie expérimentale*. Paris: Imprimerie royale, 1843.

Coleman, William, and Frederic L. Holmes (eds.), *The Investigative Enterprise: Experimental Physiology in Nineteenth-Century Medicine*. Berkeley: University of California Press, 1988.

Conrad, Sebastian, *What Is Global History?*. Princeton: Princeton University Press, 2016.

Corfield, Penelope J., 'From Poison Peddlers to Civic Worthies: The Reputation of the Apothecaries in Georgian England', *Social History of Medicine*, 22 (1) (2009): 1–21.

Costello, Francis J., *Enduring the Most: The Biography of Terence MacSwiney*. Dingle, Ireland: Brandon Books, 1996.

Counihan, Carole, and Penny van Esterik (eds.), *Food and Culture: A Reader*. London: Routledge, 1997.

Cunningham, Andrew, Perry Williams, Andrew Wear, and Roger French (eds.), *The Laboratory Revolution in Medicine*. Cambridge: Cambridge University Press, 1992.

DAR, *Da Mosè a Succi. Storia e fisiologia del digiuno*. Florence: Fieramosca, 1886.

Daston, Lorraine (ed.), *Things that Talk: Object Lessons from Art and Science*. New York: Zone Books, 2004.

Daston, Lorraine, and Peter Galison, *Objectivity*. New York: Zone Books, 2010.

Davis, Natalie Zemon, 'Decentering History: Local Stories and Cultural Crossings in a Global World', *History and Theory*, 50 (2) (2011): 188–202.

Dazinger, Kurt, 'Wilhelm Wundt and the Emergence of Experimental Psychology', in Robert Olby et al. (eds.), *Companion to the History of Modern Science*. London: Routledge, 1990, 396–409.

Diezemann, Nina, *Die Kunst des Hungerns. Essstörungen in Literatur und Medizin um 1900*. Berlin: Kadmos, 2006.

Digby, Anne, *Making a Medical Living: Doctors and Patients in the English Market for Medicine, 1720–1911*. Cambridge: Cambridge University Press, 1994.

Dinges, Martin, et al. (eds.), *Medical Practice, 1600–1900: Physicians and Their Patients*. Leiden: Brill, 2016.

Dreifuss, Jean Jacques, 'Charles Chossat (1796–1875), physiologiste, médecin et homme politique genevois', *Gesnerus*, 45 (1988): 239–261.

Dubins, Barbara, 'Nineteenth-Century Travel Literature on the Comoro Islands: A Bibliographical Essay', *African Studies Bulletin*, 12 (2) (1969): 138–146.

Duffy, John, *From Humors to Medical Science: A History of American Medicine*. Urbana: University of Illinois Press, 1993.

Dyck, Erika, and Larry Stewart, *The Uses of Humans in Experiment: Perspectives from the Seventeenth to the Twentieth Century*. Amsterdam: Clio Medica, 2016.

Eddy, Matthew D., 'The Sparkling Nectar of Spas; or, Mineral Water as a Medically Commodifiable Material in the Province, 1770–1805', in Ursula Klein and Emma Spary (eds.), *Materials and Expertise in Early Modern Europe: Between Market and Laboratory*. Chicago: University of Chicago Press, 2010, 198–224.

Edelman, Nicole, *Histoire de la voyance et du paranormal. Du XVIIIe siècle à nos jours*. Paris: Seuil, 2006.

Ehret, Arnold, *Rational Fasting for Physical, Mental and Spiritual Rejuvenation*. Los Angeles: Health Center of Los Angeles, 1926.

Elliot, Paul, 'Vivisection and the Emergence of Experimental Physiology in Nineteenth-Century France', in Rupke (ed.), *Vivisection in Historical Perspective*, 48–77.

Ellmann, Maud, *The Hunger Artists: Starving, Writing and Imprisonment*. London: Virago Press, 1993.

Figuier, Louis, 'Les nouveaux jeûneurs. Succi et Merlatti', *L'Année scientifique et industrielle*, 30 (1886): 395–404.

Filippi, Angiolo, 'Il Sor Giovanni Succi. Digiunatore e l'Academia Medico Fisica Fiorentina', *Lo Sperimentale*, 61 (1888): 324–333, 407–428.

Finkenstein, Gabriel, *Émile du Bois-Reymond: Neuroscience, Self, and Society in Nineteenth-Century Germany*. Cambridge, Mass.: MIT Press, 2013.

Flandrin, Jean-Louis, and Massimo Montanari (eds.), *Food: A Culinary History from Antiquity to the Present*. New York: Columbia University Press, 1999.

Flint, Austin, *A Text-book of Human Physiology*, 4th ed. London: H. K. Lewis, 1888.

Fonvielle, Wilfrid de, *Les endormeurs. La vérité sur les hypnotisants, les suggestionistes, les magnétiseurs, les donatistes, les braïdistes*. Paris: Librairie illustrée, 1887.

Histoire de la navigation aérienne. Paris: Hachette, 1911.

Mort de faim. Étude sur les nouveaux jeûneurs. Paris: Librairie illustrée, 1886.

Les saltimbanques de la science. Comment ils font des miracles. Paris: M. Dreyfous, 1884.

Fonvielle, Wilfrid de, James Glaisher, Camille Flammarion, and Gaston Tissandier, *Voyages aériens ouvrage contenant 117 gravures*. Paris: Hachette, 1870.

Fowler, Robert, *A Complete History of the Welsh Fasting Girl Sarah Jacob, with Comments Thereon; and Observations on Death from Starvation*. London, 1871.

Fox, Robert, *The Savant and the State: Science and Cultural Politics in Nineteenth-Century France*. Baltimore: Johns Hopkins University Press, 2012.

Fritzsche, Peter, *Reading Berlin 1900*. Cambridge, Mass.: Harvard University Press, 1996.

Fyfe, Aileen, and Bernard Lightman (eds.), *Science in the Marketplace: Nineteenth-Century Sites and Experiences*. Chicago: University of Chicago Press, 2007.

Garampazzi, Carlo, and Giuseppe Raineri, *Il digiuno de G. Succi. Considerazioni fisico-patologiche*. Pallanza and Rome: Eredi Vercellini, 1886.

Geddes, Jennian F., 'Culpable Complicity: The Medical Profession and the Forcible Feeding of Suffragettes, 1909–1914', *Women's History Review*, 17 (1) (2008): 79–94.

Geison, Gerald L., *Michael Foster and the Cambridge School of Physiology: The Scientific Enterprise in Late Victorian Society*. Princeton, N.J.: Princeton University Press, 1978.

(ed.), *Physiology in the American Context, 1850–1940*. Bethesda, Md.: American Physiological Society, 1987.

Gentilcore, David, *Medical Charlatanism in Early Modern Italy*. Oxford: Oxford University Press, 2006.

Gieryn, Thomas F., 'Boundary-Work and the Demarcation of Science from Non-Science: Strains and Interests in Professional Ideologies of Scientists', *American Sociological Review*, 48 (6) (1983): 781–795.

Cultural Boundaries of Science: Credibility on the Line. Chicago: University of Chicago Press, 1999.

Glaser, Catherine, 'Journalisme et critique scientifiques. L'exemple de Victor Meunier', *Romantisme*, 65 (1989): 27–48.

Golan, Tal, 'Blood Will Out: Distinguishing Humans from Animals and Scientists from Charlatans in the Nineteenth-Century American Courtroom', *Historical Studies in the Physical and Biological Sciences*, 31 (1) (2000): 93–124.

Gooldin, Sigal, 'Fasting Women, Living Skeletons and Hunger Artists: Spectacles of Body and Miracles at the Turn of a Century', *Body and Society*, 9 (2) (2003): 27–53.

Gould, George M., and Walter L. Pyle, *Anomalies and Curiosities of Medicine*. Philadelphia: W. B. Saunders, 1897.

Grant, Kevin, 'Fearing the Danger Point: The Study and Treatment of Human Starvation in the United Kingdom and India, c. 1880–1974', in McCue (ed.), *Comparative Physiology of Fasting, Starvation, and Food Limitation*, 365–378.

Graus, Andrea, *Ciencia y espiritismo en España (1880–1930)*. Granada: Comares, 2019.

Greenhalgh, Paul, *Ephemeral Vistas: The Expositions Universelles, Great Exhibitions and World's Fairs, 1851–1939*. Manchester: Manchester University Press, 1988.

Gregory, Frederick, *Scientific Materialism in Nineteenth-Century Germany*. Dordrecht: Reidel, 1977.

Griffith, R. Marie, 'Apostles of Abstinence: Fasting and Masculinity during the Progressive Era', *American Quarterly*, 52 (4) (2000): 599–638.

Grimsley-Smith, Melinda, 'Revisiting a "Demographic Freak": Irish Asylums and Hidden Hunger', *Social History of Medicine*, 25 (2) (2011): 307–323.

Gronau, Barbara, 'Asceticism Poses a Threat: The Enactment of Voluntary Hunger', in Lagaay and Lorber (eds.), *Destruction in the Performative*, 99–109.

Gronau, Barbara, and Alice Lagaay (eds.), *Ökonomien der Zurückhaltung. Kulturelles Handeln zwischen Askese und Religion*. Bielefeld: Transcript Verlag, 2010.

Guarneri, Patrizia, 'Moritz Schiff (1823–1896): Experimental Physiology and Noble Sentiment in Florence', in Rupke (ed.), *Vivisection in Historical Perspective*, 105–124.

Guerrini, Anita, *Experimenting with Humans and Animals: From Galen to Animal Rights*. Baltimore: Johns Hopkins University Press, 2003.

Gunn, Robert A., *40 Days without Food: A Biography of Henry S. Tanner MD, Including a Complete and Accurate History of His Wonderful Fast, 42 Days in Minneapolis, and 40 Days in New York City, with Valuable Deductions*. New York: Albert Metz, 1880.

Habermas, Tilma, and A. Beveridge, 'Historical Continuities and Discontinuities between Religious and Medical Interpretations of Extreme Fasting: The Background to Giovanni Brugholi's Description of Two Cases of Anorexia Nervosa in 1875', *History of Psychiatry*, 3 (1992): 431–455.

Hagner, Michael, 'Scientific Medicine', in David Cahan (ed.), *From Natural Philosophy to the Sciences: Writing the History of Nineteenth-Century Science*. Chicago: University of Chicago Press, 2003, 49–87.

Hahn, Hazel, *Scenes of Parisian Modernity: Culture and Consumption in the Nineteenth Century*. London: Palgrave Macmillan, 2009.

Haller, John S., *The History of American Homeopathy: From Rational Medicine to Holistic Health Care*. New Brunswick, N.J.: Rutgers University Press, 2009.

Medical Protestants: The Eclectics in American Medicine, 1825–1939. Carbondale: Southern Illinois University Press, 1994.

Hamlin, Christopher, 'Chemistry, Medicine, and the Legitimization of English Spas, 1740–1840', *Medical History*, 10 (1990): 67–81.

A Science of Impurity: Water Analysis in Nineteenth-Century Britain. Bristol: Adam Hilger, 1990.

Hammond, William A., *Fasting Girls: Their Physiology and Pathology*. New York: G. P. Putnam's Sons, 1879.

Hazzard, Linda B., *Scientific Fasting: The Ancient and Modern Key to Health*. New York: Grant Publications, 1927.

Hilgartner, Stephen, 'The Dominant View of Popularisation: Conceptual Problems, Political Issues', *Social Studies of Science*, 20 (1990): 519–539.

Hochadel, Oliver, and Agustí Nieto-Galan (eds.), *Barcelona: An Urban History of Science and Modernity (1888–1929)*. London: Routledge, 2016.

(eds.), *Urban Histories of Science: Making Knowledge in the City, 1820–1940.* London: Routledge, 2019.

Hodgson, Barbara, *In the Arms of Morpheus: The Tragic History of Laudanum, Morphine, and Patent Medicines*. Buffalo, N.Y.: Firefly Books, 2001.

Holmes, Frederick L., 'The Intake–Output Method of Quantification in Physiology', *Historical Studies in the Physical and Biological Sciences*, 17 (2) (1987): 235–270.

Hu, Aiqun, 'The Global Insurance Movement since the 1880s', *Journal of Global History*, 5 (2010): 125–148.

Human Wonder of the World, The. The Celebrated Dr Tanner. A Full Account of his Forty Days Fast. Philadelphia: Barclay and Co., 1880.

Jütte, Robert, 'The Historiography of Nonconventional Medicine in Germany: A Concise Overview', *Medical History*, 43 (1999): 342–358.

'The Paradox of Professionalism: Homeopathy and Hydropathy as Unorthodoxy in Germany in the Nineteenth and Early Twentieth Century', in Robert Jütte, Guenter B. Risse, and John Woodward (eds.), *Culture, Knowledge and Healing: Historical Perspectives of Homeopathic Medicine in Europe and North America*. Sheffield: EAHMH, 1998, 65–68.

Jütte, Robert, Guenter B. Risse, and John Woodward (eds.), *Culture, Knowledge and Healing: Historical Perspectives of Homeopathic Medicine in Europe and North America*. Sheffield: EAHMH, 1998.

Kafka, Franz, *A Hunger Artist and Other Stories: Translated by Joyce Crick, with an Introduction and Notes by Ritchie Roberts*. Oxford: Oxford University Press, 2012 [1922].

Kamminga, Harmke, and Andrew Cunningham (eds.), *The Science and Culture of Nutrition, 1840–1940.* Amsterdam: Rodopi, 1995.

Klein, Ursula, and Emma C. Spary, *Materials and Expertise in Early Modern Europe: Between Market and Laboratory*. Chicago: University of Chicago Press, 2009.

Kremer, Richard L., 'Physiology', in Peter Bowler and John V. Pickstone (eds.), *The Cambridge History of Science*, vol. VI, *The Modern Biological and Earth Sciences*. Cambridge: Cambridge University Press, 2009, 342–366.

Labbé, Henri, 'L'inanition. Ses aspects physiologiques et sociaux', *Revue scientifique*, 18 (1908), 546–552.

Labisch, Alfons, *Homo Hygienicus. Gesundheit und Medizin in der Neuzeit.* Frankfurt: Campus Verlag, 1992.

Laborie, Léonard, 'Global Commerce in Small Boxes: Parcel Post, 1878–1913', *Journal of Global History*, 10 (2015): 235–258.

Lachapelle, Sofie, 'Between Miracle and Sickness: Louise Lateau and the Experience of Stigmata and Ecstasy', *Configurations*, 12 (2004): 77–105.

Investigating the Supernatural: From Spiritism and Occultism to Psychical Research and Metapsychics in France, 1853–1931. Baltimore: Johns Hopkins University Press, 2011.

Lagaay, Alice, and Michael Lorber (eds.), *Destruction in the Performative.* Amsterdam: Rodopi, 2012.

Lange-Kirchheim, Astrid, 'Nachrichten vom italienischen Hungerkünstler Giovanni Succi. Neue Materialien zu Kafkas "Hungerkünstler"', in Johannes Cremerius, et al. (eds.), *Größenphantasien, Freiburger literaturpsychologische Gespräche*. Würzburg: Königshausen und Neumann, 1999, 315–340.

Lassignardie, Henri, *Essai sur l'état mental dans l'abstinence*. Bordeaux: Imprimerie du Midi, 1897.

Leakey, Thomas H., *A History of Psychology: Main Currents in Physiological Thought*. Englewood Cliffs, N.J.: Prentice-Hall, 1992.

Lederer, Susan, *Subjected to Science: Human Experimentation in America before the Second World War*. Baltimore: Johns Hopkins University Press, 1995.

Lenoir, Timothy, 'Laboratories, Medicine and Public Life in Germany, 1830–1849: Ideological Roots of the Institutional Revolution', in Andrew Cunningham, et al. (eds.), *The Laboratory Revolution in Medicine*. Cambridge: Cambridge University Press, 1992, 14–71.

'Science for the Clinic: Science Policy and the Formation of Carl Ludwig's Institute in Leipzig', in William Coleman and Frederic L. Holmes (eds.), *The Investigative Enterprise: Experimental Physiology in Nineteenth-Century Medicine*. Berkeley: University of California Press, 1988, 139–178.

The Strategy of Life: Teleology and Mechanics in Nineteenth-Century German Biology. Dordrecht: Reidel, 1982.

Lesch, John E., *Science and Medicine in France: The Emergence of Experimental Physiology, 1890–1855*. Cambridge, Mass.: Harvard University Press, 1984.

Lignot, Jean-Hervé, and Yvon LeMaho, 'A History of Modern Research into Fasting, Starvation and Inanition', in McCue (ed.), *Comparative Physiology of Fasting, Starvation, and Food Limitation*, 7–23.

Livingstone, David, *Putting Science in Its Place: Geographies of Scientific Knowledge*. Chicago: University of Chicago Press, 2003.

Livingstone, David, and Charles W. Withers (eds.), *Geographies of Nineteenth-Century Science*. Chicago: University of Chicago Press, 2011.

Loye, Paul, *La mort par decapitation*. Paris: Publications du progrès médicale, 1888.

Luciani, Luigi, *Il cervelletto. Nuovi studi di fisiologia normale e patologica*. Florence: Coi tipi dei successori Le Monnier, 1891.

Fisiologia del digiuno: studi sull'uomo. Florence: Tipografia dei Successori Le Monnier, 1889.

Fisiologia dell'uomo, 3 vols. Milan: Societa Editrice Libraria, 1901.

Human Physiology, 4 vols. London: Macmillan and Co. Ltd, 1911.

Das Hungern. Studien und Experimente am Menschen. Hamburg and Leipzig: Leopold Voss, 1890.

'Sullo Stato generale del Succi durante il suo digiuno di trenta giorni. Prima Communicazione Fatta all'Accademia Médico-Fisica Fiorentina nella seduta pubblica del 15 Aprile del 1888', *Lo Sperimentale*, 61, (1888): 424–439.

Tratado didáctico de fisiologia humana. Barcelona: Antonio Virgili, 1901.

Ludwig, Carl, *Lehrbuch der Physiologie des Menschen*. Heidelberg: Winter, 1852.

Lusk, Graham, *Elements of the Science of Nutrition*. Philadelphia: W. B. Saunders, Co., 1906.

McCue, Marshall D. (ed.), *Comparative Physiology of Fasting, Starvation, and Food Limitation*. Dordrecht: Springer, 2012.

Mackaman, Douglas P., *Leisure Settings: Bourgeois Culture, Medicine, and the Spa in Modern France*. Chicago: University of Chicago Press, 1998.

Maglioni, Luis C., *Mis 37 días de ayuno (auto-experiencia). A propósito de los casos de Mr. MacSwiney y sus 11 compañeros, ayunadores políticos en las prisiones de Brixton y de Cork. 'Huelga de Hambre'*. Buenos Aires: Imprenta San Martín, 1920.

Manni, Ermanno, and Laura Petrosini, 'Luciani's Work on the Cerebellum, a Century Later', *Trends in Neurosciences*, 20 (1997): 112–116.

Markowitz, Gerald E., and David Rosner, *Dying for Work: Workers' Safety and Health in Twentieth-Century America*. Bloomington: Indiana University Press, 1989.

Marks, Lara (ed.), *Useful Bodies: Humans in the Service of Medical Science in the Twentieth Century*. Baltimore: Johns Hopkins University Press, 2003.

Markus, Thomas A., *Buildings and Power: Freedom and Control in the Origin of Modern Building Types*. London: Routledge, 1993.

Martín, Avelino, *El ayuno de Succi. Contribución al estudio de la inanición*. Barcelona: Tipografía de la Casa Provincial de Caridad, 1889.

Martínez de Velasco, Eusebio, 'El viajero italiano Juan Succi en el día vigesimoctavo de su ayuno', *La Ilustración Española y Americana*, 36 (1886): 182.

Mavropoulos, Nikolaos, 'The Japanese Expansionism in Asia and the Italian Expansion in Africa: A Comparative Study of the Early Italian and Japanese Colonialism', PhD. Rome: La Sapienza, 2019.

Mazower, Mark, *Dark Continent: Europe's Twentieth Century*. London: Allen Lane, 1998.

Miller, Ian, *A Modern History of the Stomach: Gastric Illness, Medicine and British Society, 1800–1950*. London: Pickering & Chatto, 2011.

'A Prostitution of the Profession? Forcible Feeding, Prison Doctors and the British State, 1909–1914', *Social History of Medicine*, 26 (2) (2013): 225–245.

Mitchell, Breon, 'Kafka and the Hunger Artists', in Alan Udoff (ed.), *Kafka and the Contemporary Critical Performance: Centenary Readings*. Bloomington: Indiana University Press, 1987, 236–255.

Moleschott, Jacob, *Dell'alimentazione. Trattato popolare*. Milan: Treves, 1871.

Monin, Ernest, *L'hygiène de l'estomac. Guide pratique de l'alimentation*. Paris: O. Doin, n.d.

L'hygiène du travail. Guide médical des industries et professions. Paris: J. Hetzel et Cie, 1889.

Le propos du docteur. Médecine sociale, hygiène générale à l'usage des gens du monde. Paris: E. Giraud et Cie, 1885.

Monin, Ernest, and Philippe Maréchal, *Stefano Merlatti. Histoire d'un jeûne célèbre, précédée d'une étude anecdotique, physiologique et médicale sur le jeûne et les jeûneurs*. Paris: C. Marpon et E. Flammarion, 1887.

Monroe, John Warne, *Laboratories of Faith: Mesmerism, Spiritism and Occultism in Modern France*. Ithaca: Cornell University Press, 2008.

Moody, Alys, *The Art of Hunger: Aesthetic Autonomy and the Afterlives of Modernism*. Oxford: Oxford University Press, 2018.

Morabito, Carmela, 'Luigi Luciani and the Localization of Brain Functions: Italian Research within the Context of European Neurophysiology at the

End of the Nineteenth Century', *Journal of the History of the Neurosciences*, 9 (2000): 180–200.
Morgulis, Sergius, *Fasting and Undernutrition: A Biological and Sociological Study of Inanition*. New York: Dutton and Company, 1923.
Morus, Iwan Rhys, *Frankenstein's Children: Electricity, Exhibition, and Experiment in Early Nineteenth-Century London*. Princeton, N.J.: Princeton University Press, 1998.
Mosley, Adam, 'Objects, Texts and Images in the History of Science', *Studies in History and Philosophy of Science*, 38 (2007): 289–302.
Mosso, Angelo, *Fatigue*. London: Swan Sonnenschein & Co. Ltd, 1906.
Mulberger, Annette, 'Wundt Contested: The First Crisis Declaration in Psychology', *Studies in History and Philosophy of Science Part C: Studies in History and Philosophy of Biological and Biomedical Sciences*, 43 (2012): 434–444.
Mulberger, Annette, and Mònica Balltondre, 'Metapsychics in Spain: Acknowledging or Questioning the Marvellous?', *History of Human Sciences*, 25 (4) (2012): 108–130.
Neswald, Elizabeth, 'Food Fights: Human Experiments in Late Nineteenth-Century Nutrition Physiology', *Clio Medica*, 95 (2016): 170–193.
'Francis Gano Benedict's Reports of Visits to Foreign Laboratories and the Carnegie Nutrition Laboratory', *Actes d'Història de la Ciència i de la Tècnica, (nova epoca)*, 4 (2011): 11–32.
'Strategies of International Community-Building in Early Twentieth-Century Metabolism Research: The Foreign Laboratory Visits of Francis Gano Benedict', *Historical Studies in the Natural Sciences*, 43 (1) (2013): 1–40.
Neswald, Elizabeth, and David F. Smith, *Setting Nutritional Standards: Theory, Policies, Practices*. Rochester, N.Y.: Rochester Studies in Medical History, 2017.
Nicholson, Virginia, *Among the Bohemians: Experiments in Living 1900–1939*. London: Penguin, 2003.
Nickell, Joe, 'Mystery of Mollie Fancher: The Fasting Girl and Others Who Lived without Eating', *Skeptical Inquirer*, 41 (6) (2017): 18–21.
Nieto-Galan, Agustí, 'Antonio Gramsci Revisited: Historians of Science, Intellectuals, and the Struggle for Hegemony', *History of Science*, 46 (2011): 453–478.
'Historicitat i heterodòxia. La divulgació científica de Louis Figuier (1819–1894)', *Quaderns de Filologia. Estudis Lingüístics*, 17 (2012): 81–94.
'Hunger Artists and Experimental Physiology in the Late Nineteenth Century: Mr Giovanni Succi Meets Dr Luigi Luciani in Florence', *Social History of Medicine*, 28 (1) (2015): 82–107.
'The Images of Science in Modern Spain: Rethinking the "Polémica"', in Kostas Gavroglu (ed.), *The Sciences in the European Periphery during the Enlightenment*. Dordrecht: Kluwer, 1998, 65–86.
Science in the Public Sphere: A History of Lay Knowledge and Expertise. London: Routledge, 2016.
'Scientific "Marvels" in the Public Sphere: Barcelona and Its 1888 International Exhibition', *HoST – Journal of History of Science and Technology*, 6 (2012): 33–63.

'Useful Charlatans: The Fasting Contest of Giovanni Succi and Stefano Merlatti in Paris, 1886', *Science in Context*, 33 (4) (2020): 405–422.

Noakes, Richard, *Physics and Psychics: The Occult and the Sciences in Modern Britain*. Cambridge: Cambridge University Press, 2019.

Offer, Avner, 'Body Weight and Self Control in the US and Britain since the 1950s', *Social History of Medicine*, 14 (1) (2001): 79–106.

Onoranza a Luigi Luciani, nel XXV anniversario del suo insegnamento. Ascoli Piceno. 29 Aprile 1900. Ascoli Piceno: Tip. Cesari, 1900.

Orosi, Giuseppe, *Manuale dei medicamenti galenici e chimici con la descrizione dei loro caratteri, la loro preparazione, la virtù terapeutica, le formule di uso medico, le incompatibilità relative, le adulterazioni commerciali, gli antidoti ec*. Florence: Eugenio e F. Cammelli, 1867.

Pacifici, Paola, 'Le corps. Anatomie d'un symbole', *Protée*, 36 (1) (2008): 29–38.

Parot, Françoise, and Marc Richelle, *Introduction à la psychologie. Histoire et méthodes*. Paris: Presses Universitaires de France, 1992.

Payer, Peter, 'Hungerkünstler. Anthropologisches Experiment und modische Sensation', in Brigitte Felderer and Ernst Strouhal (eds.), *Rare Künste. Zur Kultur- und Mediengeschichte der Zauberkunst*. Vienna: Springer, 2006, 254–268.

Hungerkünstler eine verschwundene Attraktion. Vienna: Sonderzahl, 2000.

Hungerkünstler in Wien. Eine verschwundene Attraktion. Vienna: Verlag Sonderzahl, 2002.

Person, Jutta, 'Abnormität und Irrsinn – Das Spektakel des Hungerkünstlers Succi', in Torsten Hahn, Jutta Person, and Nicolas Pethes (eds.), *Grenzgänge zwischen Wahn und Wissen. Zur Koevolution von Experiment und Paranoia 1850–1910*, Frankfurt am Main: Campus, 2002, 240–254.

Pickstone, John V., 'Physiology and Experimental Medicine', in Robert Olby, et al. (eds.), *Companion to the History of Modern Science*. London: Routledge, 1990, 728–742.

Piqué, Joel, 'Procesos de construcción social y científica de la homeopatía en Catalunya (1890–1924)', PhD Thesis, Universitat Autònoma de Barcelona, Bellaterra, 2018.

Plas, Régine, 'Psychology and Psychical Research in France around the End of the Nineteenth Century', *History of the Human Sciences*, 25 (2) (2012): 91–107.

Podgorny, Irina (ed.), *Charlatanes. Crónicas de remedios incurables*. Buenos Aires: Eterna Cadencia Editora, 2012.

'The Elk, the Ass, the Tapir, Their Hooves, and the Falling Sickness: A Story of Substitution and Animal Medical Substances', *Journal of Global History*, 13 (2018): 46–68.

Podgorny, Irina, and Daniel Gethmann, '"Please, Come In": Being a Charlatan, or the Question of Trustworthy Knowledge', *Science in Context*, 33 (4) (2020): 355–361.

Pogliano, Claudio, 'La fisiologia in Italia fra ottocento e novecento', *Nuncius*, 6 (1) (1991): 97–121.

Porter, Roy, *Quacks, Fakers and Charlatans in English Medicine*. Stroud: Tempus, 2000.

Porter, Roy, and Mikulas Teich (eds.), *Drugs and Narcotics in History*. Cambridge: Cambridge University Press, 1995.

Priori, Domenico, 'Il digiunanote e lo scienziato', *Rendiconti Accademia Nazionale delle Scienze detta dei XL. Memorie di Scienze Fisiche e Naturali*, 135 (2017): 1–9.

Rabinach, Anson, *The Human Motor: Energy, Fatigue and the Origins of Modernity*. Berkeley: University of California Press, 1992.

Raj, Kapil, 'Go-Betweens, Travelers, and Cultural Translators', in Bernard Lightman (ed.), *A Companion to the History of Science*. Chichester: Wiley-Blackwell, 2016, 39–57.

Ramsey, Matthew, 'Sous le régime de la législation de 1803. Trois enquêtes sur les charlatans au XIXe siècle', *Revue d'histoire moderne et contemporaine*, 27 (1980): 485–500.

Rees, John, *Textbook of Medical Jurisprudence and Toxicology*, 4th ed. Philadelphia: Blakiston's Son and Co., 1895.

Rheinberger, Hans-Jörg, and Michael Hagner (eds.), *Die Experimentalisierung des Lebens. Experimentalsysteme in den Biologischen Wissenschaften, 1850–1950*. Berlin: Akademie Verlag, 1993.

Richards, Graham, *Putting Psychology in Its Place: Critical Historical Perspectives*, 3rd ed. London: Routledge, 2010.

Richards, Thomas, *The Commodity Culture of Victorian England: Advertising and Spectacle, 1851–1914*. Stanford: Stanford University Press, 1990.

Richet, Charles (ed.), *Dictionnaire de physiologie*, 10 vols. Paris: Alcan, 1895–1928.

'Physiologie. Course de physiologie de la Faculté de Médecine de Paris. L'inanition chez les animaux', *Revue scientifique*, 21 (1889): 641–647; 23 (1889): 711–715.

'Physiologie. Course de physiologie de la Faculté de Médecine de Paris. L'inanition chez l'homme', *Revue scientifique*, 4 (1889): 106–112; 26 (1889): 801–805.

Physiologie. Travaux du laboratoire de M. Charles Richet. Chimie physiologique. Toxicologie. Paris: F. Alcan, 1893–1895.

Rochas, Albert de, *La suspension de la vie*. Paris: Dorbon-ainé, 1913.

Rupke, Nicolaas A. (ed.), *Vivisection in Historical Perspective*. London: Croom Helm, 1987.

Russell, Sharman Apt, *Hunger: An Unnatural History*. New York: Basic Books, 2005.

Saltarino, Signor, *Fahrend Volk. Abnormalitäten, Kuriositäten und interessante Vertreter der wanderen Künstlerwelt*. Leipzig: J. J. Weber, 1895.

Sastre-Juan, Jaume, and Jaume Valentines, 'Technological Fun: The Politics and Geographies of Amusement Parks', in Hochadel and Nieto-Galan (eds.), *Barcelona: An Urban History of Science and Modernity*, 92–112.

Schütz, Mathias, 'After Pettenkofer: Munich's Institute of Hygiene and the Long Shadow of National Socialism, 1894–1974', *International Journal of Medical Microbiology*, 310 (5) (2020): 1–7.

Secord, James, 'Knowledge in Transit', *Isis*, 95 (2004): 654–672.

Shapin, Steven, '"You Are What You Eat": Historical Changes in Ideas about Food and Identity', *Historical Research*, 87 (2014): 377–392.

Sharp, Lynn L., *Secular Spirituality: Reincarnation and Spiritism in Nineteenth-Century France*. Plymouth: Lexington Books, 2006.

Sherson, Erroll, *London's Lost Theatres of the Nineteenth Century*. London: John Lane, 1925.

Shinn, Terry, and Richard Whitley (eds.), *Expository Science: Forms and Functions of Popularization*. Dordrecht: Reidel, 1985.

Smith, Roger, 'The Physiology of the Will: Mind, Body, and Psychology in the Periodical Literature, 1855–1875', in Geoffrey N. Cantor and Sally Shuttleworth (eds.), *Science Serialized: Representations of the Sciences in Nineteenth-Century Periodicals*. Cambridge, Mass.: MIT Press, 2004, 81–110.

Spary, Emma C., *Eating the Enlightenment: Food and the Sciences in Paris*. Chicago: University of Chicago Press, 2012.

Spary, Emma C., and Anya Zilberstein (eds.), 'Food Matters', *Osiris*, 35 (2020).

Spary, Emma C., and Anya Zilberstein, 'On the Virtues of Historical Entomophagy', *Osiris*, 35 (2020): 1–19.

Spillane, Joseph, 'The Making of an Underground Market: Drug Selling in Chicago, 1900–1940', *Journal of Social History*, 32 (1998): 27–47.

Stacey, Michelle, *Fasting Girl: A True Victorian Medical Mystery*. New York: J. P. Tarcher/ Putnam, 2002.

Succi, Giovanni, *Commercio in Africa. Il Madagascar, l'isola di Johanna e l'arcipelago di Comoro, Zanzibar e Mozambese*. Milan: Tipografia Nazionale, 1881.

Swartz Rose, Mary, *The Foundations of Nutrition*. New York: Macmillan Company, 1927.

Tanner, Henry S., *The Human Body: A Volume of Divine Revelations, Governed by Laws of God's Ordaining, Equally with Those Written on 'Tables of Stone' or in the Bible*. Long Beach, Calif., 1908.

Todes, Daniel P., 'Pavlov and the Bolsheviks', *History and Philosophy of Life Sciences*, 17 (1995): 379–418.

Train, George Francis, *My Life in Many States and in Foreign Lands*. New York: D. Appleton and Company, 1902.

Treitel, Corinna, *Eating Nature in Modern Germany: Food, Agriculture and Environment, c. 1870 to 2000*. Cambridge: Cambridge University Press, 2017.

'Max Rubner and the Biopolitics of Rational Nutrition', *Central European History*, 41 (2008): 1–25.

'Nutritional Modernity: The German Case', *Osiris*, 35 (1) (2020): 183–203.

Trompp, Marlene (ed.), *Victorian Freaks: The Social Context of Freakery in Britain*. Columbus: Ohio State University Press, 2008.

Tucker, Richard P., Tait Keller, John McNeill, and Martin Schmid, *Environmental Histories of the First World War*. Cambridge: Cambridge University Press, 2018.

Tucker, Todd, *The Great Starvation Experiment: The Heroic Men Who Starved So That Millions Could Live*. New York: Free Press, 2006.

Turner, Bryan S., 'The Government of the Body: Medical Regimens and the Rationalization of Diet', *British Journal of Sociology*, 33 (2) (1982): 254–269.

Udoff, Allan (ed.), *Kafka and the Contemporary Critical Performance: Centenary Readings*. Bloomington: Indiana University Press, 1987.

Van Deth, Ron, R. Meermann, and Walter Vandereycken, *Hungerkünstler, Fastenwunder, Magersucht. Eine Kulturgeschichte der Ess-Störungen*. Zülpich: Biermann, 1990.

Van Deth, Ron, and Walter Vandereycken, 'The Striking Age-Old Minority of Fasting Males in the History of Anorexia Nervosa', *Food and Foodways*, 7 (2) (1997): 119–130.

Van Strien, Marij, 'Vital Instability: Life and Free Will in Physics and Physiology, 1860–1880', *Annals of Science*, 72 (3) (2015): 381–400.

Vandereycken, Walter, and Ron van Deth, *From Fasting Saints to Anorexic Girls: The History of Self-Starvation*. London: Athlone Press, 1994.

Vargas, Joel, 'Calibrar la alimentación. La estandarización del calorímetro en México', in Laura Cházaro, Miruna Achim, and Nuria Valverde (eds.), *Piedra, papel y tijera. Instrumentos en las ciencias en México*. Mexico City: UAM Cuajimalpa, 2018, 211–244.

Vernon, James, *Hunger: A Modern History*. Cambridge, Mass.: Belknap Press, 2007.

Weisz, George, 'The Emergence of Medical Specialization in the Nineteenth Century', *Bulletin of the History of Medicine*, 77 (2003): 536–575.

'Water Cures and Science: The French Academy of Medicine and Mineral Waters in the Nineteenth Century', *Bulletin of the History of Medicine*, 64 (3) (1990): 393–416.

Williams, Elizabeth A., *Appetite and Its Discontents: Science, Medicine and the Urge to Eat, 1750–1950*. Chicago: University of Chicago Press, 2020.

'Neuroses of the Stomach: Eating, Gender, and Psychopathology in French Medicine, 1800–1870', *Isis*, 98 (2007): 54–79.

The Physical and the Moral: Anthropology, Physiology, and Philosophical Medicine in France, 1750–1850. Cambridge: Cambridge University Press, 1994.

Winchester, Simon, *Krakatoa: The Day the World Exploded, 27 August 1883*. London: Viking, 2003.

Wolf, Stewart, *Brain, Mind, and Medicine: Charles Richet and the Origins of Physiological Psychology*. London: Transaction Publishers, 2012.

Wolffram, Heather, 'Hallucination or Materialization? The Animism versus Spiritism Debate in Late Nineteenth-Century Germany', *Journal of the History of the Human Sciences*, 25 (2) (2012): 45–66.

Index

Printed in the United States
by Baker & Taylor Publisher Services